MW00643297

Better in Every Sense

Better in Every Sense

How the New Science of Sensation Can Help You Reclaim Your Life

NORMAN FARB
ZINDEL SEGAL

Little, Brown Spark
New York Boston London

Little, Brown Spark
Hachette Book Group
1290 Avenue of the Americas, New York, NY 10104
littlebrownspark.com

First Edition: January 2024

Little, Brown Spark is an imprint of Little, Brown and Company, a division of Hachette Book Group, Inc. The Little, Brown Spark name and logo are trademarks of Hachette Book Group, Inc.

Photographs on pages 28 and 56 by Shutterstock

ISBN 9780316434430
LCCN 2023934774

Printing 1, 2023

LSC-C

Printed in the United States of America

To everyone choosing hope over fear

Contents

Better in Every Sense

Introduction

Much Madness is divinest Sense—
To a discerning Eye

—Emily Dickinson

Maya had a comfortable life. She had a steady job, went to concerts, ate the occasional taco, did yoga, and made time for friends and hobbies. Unlike the protagonists in great works of fiction, she had experienced no catastrophic divorce, addiction, or bankruptcy. Rather than some fabled redemption arc from success to obscurity and back again, Maya's life was more even-keeled. Yet she awoke each morning feeling restless and dissatisfied. She found herself thinking, "I thought life would feel better." But she didn't feel better. She felt *stuck*.

As an accomplished problem-solver in the workplace, Maya had done the self-help thing, tried to set herself on a journey of personal growth. She followed bestselling guides to self-improvement that zeroed in on behaviors that promised to make her life better. She was told things like: be persistent, celebrate victories, avoid negative thoughts, express gratitude, stay in the moment, and visualize your goals. The idea was to build up what's helpful and shed what's unhelpful, with periodic check-ins to ensure that she was "doing it right."

But nothing helped. She finished off multiple self-help guides and still felt stuck. Even getting a promotion at work stopped moving the needle on life satisfaction. In her quiet moments, Maya wondered what to do—she had followed the best advice and still felt anxious and unfulfilled. How could she "get better" when she was mired in an existential quandary with no clear path out of the mud? When your efforts to improve leave you feeling blocked and overwhelmed, the idea of "getting better" stops making sense. What does "better" even mean?

The remedy came from an unexpected source. Maya's aunt dragged her to a local knitting group. Agreeing initially out of a sense of familial obligation, Maya was not expecting much. But to her surprise, there was something about the tactile experience of knitting, the plastic needles in her hands, the metronomic sound of the needles clicking together, the yarn sliding across her fingers, and the dexterity required, that drew her in. She felt a surprising surge of pleasure at seeing the strands of yarn transform into something larger, complete, and whole. She wasn't sure how, but after each knitting class, she felt somehow lighter, free to think more creatively, her problems receding into the distance. Most importantly, in those moments of exploration, she stopped feeling stuck. She wondered why knitting felt so different from the rest of her life but wasn't sure how to make use of this experience.

That's where we come in. There is a hidden science behind why we get stuck and what to do about it. It's not usually due to a lack of problem-solving skills, so books promising to provide such skills might be barking up the wrong tree. Instead, we begin to feel trapped when we find ourselves having to solve the *same* problems over and over again. We become prisoners of our own problem-solving strategies, with no clear way to break the cycle.

And there's a danger here, backed by more than a century of research: staying stuck *inevitably* leads to breakdown. A partner stuck in "being right" wins arguments and loses relationships. A person stuck in "saving money" accumulates wealth but not happiness. Someone stuck in "being safe" avoids catastrophe at the cost of surprise and wonder. What if our hard-won habits for maintaining the status quo have become a big part of the problem? There is a solution, but it's not what you expect.

With this book, we want to teach you how to renegotiate your habitual behaviors, restore your potential for getting unstuck, and reengage with your life. These skills will empower you to move ahead in your career, your relationship, your stalled-out creative project—paradoxically, by embracing the feeling of not knowing how to fix it. Which, if you think about it, is pretty much the way the world presents itself to us—with plenty of surprises. We're conditioned to act as if the world is predictable—that *x* plus *y* ultimately will add up to *z*, and that *z* will bring happiness. We're told that if we work hard enough, we'll acquire resources and acclaim, and surely contentment will follow . . . right?

There is, of course, tremendous cultural pressure to believe in this formula, but the conditioning doesn't come from culture alone. The fact is, our brains are wired to find patterns, adapt to them, then anticipate and respond to the world in predictable ways. Much of this is helpful—essential, in fact. The problem arises when your patterns of perception, and the habits that follow, no longer align with reality, or with what you truly want.

We've spent years—decades, actually—studying the brain network in charge of finding and following patterns. We call it the House of Habit, although it's more formally known as the default mode network (DMN). It calls the shots far more than you may realize, and much of the time it allows us to glide through life far more easily than we otherwise might. But sometimes this House of Habit gets things really, really wrong, and that's when it needs to be circumvented.

Through our research—which you'll learn a great deal more about in the pages ahead—we've discovered a way of slipping around this dominant, pattern-following part of the brain to sense what's really going on and to respond without the rigidity of the preprogrammed behaviors that are the DMN's stock-in-trade. This can be incredibly liberating, and—by the way—very good for your mental health. Escaping the House of Habit means being able to find out for yourself what makes sense in your life, what needs to change, what must be endured, and what can be savored and appreciated.

From what we see around us, Maya isn't alone in her dissatisfaction. Millions of people in all walks of life are desperate to get "unstuck." The world needs this kind of help more than ever. One of the side effects of COVID-19 was the Great Resignation, the realization by millions of workers, ranging from the cleanup crew to the CEO, that their current job—and maybe any steady job—just wasn't for them. These restless people might not have been entirely sure what they wanted, but they knew what they had wasn't it.

This sort of restlessness used to be called a "midlife crisis," a term that became fashionable during the 1950s. More recently, twenty-five-year-olds began to experience the "quarter-life crisis," especially when early disappointments meant returning home to live in Mom and Dad's basement. This crisis comes not only from realizing that we may have been sold an overidealistic worldview, but also from being unable to find a better one. When our waking life is governed too much by the DMN, the trouble can take numerous forms: writer's block, career burnout, empathy fatigue, or even a loss of momentum that has no description better than "the blahs." What's the point of trying when we seem like a side character in other people's stories, grudgingly marching from cradle to grave?

But ruminating about how life isn't going the way you want doesn't get you out of your predicament. It's just piling guilt about the blahs on top of the blahs. You need a new way of experiencing life when you're stuck resenting the experience you have. All those excellent work habits and empowering habits of mind you've learned aren't going to help. You simply can't *think* your way out of this one. In fact, all that hard work you've done to make sense of the world may well have woven the web that has you feeling stuck.

OUR PROMISE TO YOU

Many programs for getting unstuck and living a happier life map out steps to get you from a place of disorganization to a place of certainty, guiding you to act or feel a particular way. As you may have guessed, we hope to do the opposite. We want to help you limit the extent to which you exploit your existing repertoire of predictable and "appropriate" behaviors—the mental habits imposed by the DMN—and find new options for engaging with the world. This means getting beyond rule following and rigid patterns of perception and action, and learning instead to forage in uncertainty, to effectively "sense before you leap."

But we're not sending you off to an ashram to find your bliss, to subvert your faculty of reason, or to turn you away from the world. We want you to flourish in the here and now, right where you are. For that, you'll need to find a sweet spot that balances *preserving* the habits that work for you with *discovering* new ways of being that are better than what came before. Escaping the tyranny of the DMN can help you make the most of two human capacities that are critical: being able to exploit what you know while still being free to explore for what you don't. The payoff will be a revitalized you and a happier life.

It's time for life to get *better*, in every sense.

1

GETTING STUCK

It's the same thing your whole life, clean up your room, stand up straight, pick up your feet . . . be nice to your sister, don't mix beer and wine ever . . . oh yeah, don't drive on the railroad tracks.

—*Groundhog Day*

GROUNDHOG DAY EVERY DAY

Shanice took a deep breath, traded in her steady gig managing a large PetSmart store, and struck out on her own. Her confidence came from noticing how many products customers kept asking for that PetSmart didn't carry. She understood the risks of walking away from a good salary, cashing in most of her savings, and signing a twelve-month lease for office space—all without having booked a single order. Still, the network of suppliers and veterinarians she'd built up as a manager offered the perfect starting point for selling her own line of custom-designed and personalized pet toys, clothing, and GPS collars. Besides, she thrived on passion, hustle, and relentless effort—exactly what's needed to launch a successful business.

For months, she did it all: hired employees, ran strategy meetings, met with manufacturers, and did anything else that was needed. One year in, though, Shanice was spending more time at the office than ever, running on adrenaline and stale coffee. Then she began to lose it. She found herself having trouble concentrating, getting irritable at home, and misplacing orders, until one day it all came to a head. Shanice found herself sitting in a stairwell crying because she had no idea what was going on or what to do about it. "I feel used up," she said, sobbing, "there's nothing left to give."

If Shanice's story sounds familiar, that's because she's by no means alone in her struggle. Surveys indicate that 28 percent of entrepreneurs experience founder burnout.[1] The same nose-to-the-grindstone habits that make entrepreneurs successful can also lead to their downfall. The really insidious thing is that, even as she hit the wall, Shanice felt as if she had no choice but to keep banging her head against it. And from her brain's perspective, she didn't. Despite clear signs of exhaustion, her brain—specifically the default mode network—highly conditioned to respond to stress by buckling down and pushing harder, refused to entertain even the possibility of changing gears or changing course.

This kind of meltdown happens not just to entrepreneurs, but to artists, teachers, doctors, engineers—anybody who pours themselves into their work. Nearly a third of adult US workers reported feelings of work-related exhaustion before the pandemic,[2] and now more than 50 percent of workers would describe themselves as experiencing burnout.[3] Success offers no protection from this feeling—we've seen it recently with world-class athletes.[4] When the deeply ingrained approach to overcoming adversity is to keep on keepin' on, taking a break can seem like failing, even when it's clearly the right thing to do.

While increasingly common today, finding ourselves in a state where the status quo seems safe but unsustainable is hardly unique to our era. One of the greatest works in all of literature, written in the fourteenth century, begins this way:

> *At the mid-point of the path through life, I found*
> *Myself lost in a wood so dark, the way*
> *Ahead was blotted out.*

Five hundred years after Dante introduced his *Divina Commedia* with those forlorn words, Danish philosopher Søren Kierkegaard used "angst" to describe a vague dread or anxiety over nothing in particular, which he thought might just be the human condition. By the end of the nineteenth century, the French word "ennui" was being used to describe a pervasive dissatisfaction with industrial life. The Germans had their own word for it: "Weltschmerz." Musician and singer-songwriter Mark Knopfler called it "industrial disease."

Whatever term we use, the shared idea is that work, safety, and some modicum of security sometimes deliver not a sense of satisfaction, accomplishment, and peace but a deep feeling of restlessness. This free-floating ennui/Weltschmerz/angst/dread/anxiety suggests that whatever we've gained or experienced is, at this moment, just not what we'd hoped for.

It seems as if the ladder leading to deeper meaning and flourishing keeps being pulled up beyond our reach. We've met our most basic needs, but where we'd hoped to find meaning, we discover emptiness. In short, we're just not happy.

MASLOW'S GLASS CEILING

In the early 1940s, the American psychologist Abraham Maslow developed a concept that's been passed along in introductory psych courses ever since. He visualized the hierarchy of human needs in the form of a pyramid, with the stuff that keeps us alive—food, shelter, warmth—as merely the foundation, necessary but by no means sufficient for producing a feeling of life satisfaction.

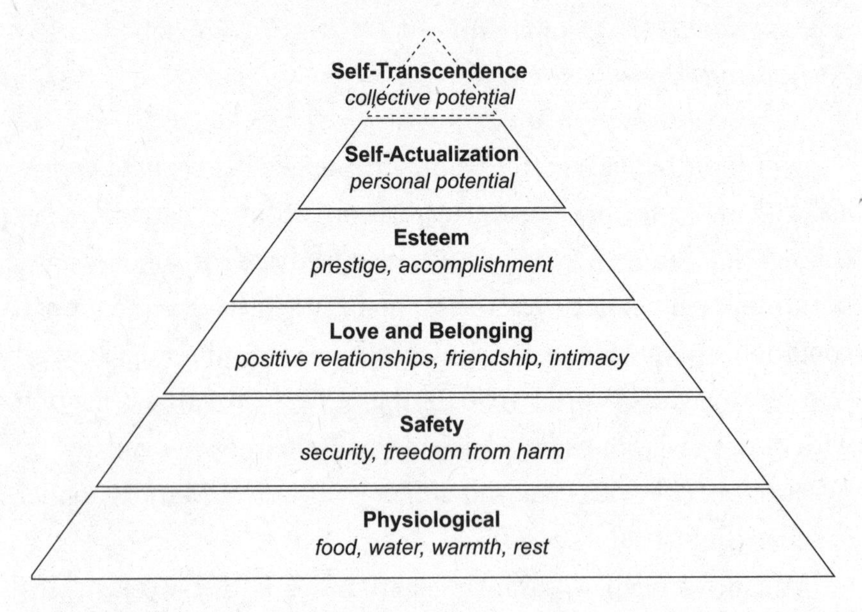

Maslow acknowledged that achieving self-actualization, or self-transcendence, is rare, with the inability to meet more basic needs serving as a "glass ceiling" on ambitions to reach the pinnacle of self-fulfillment. Complementing the burnout data discussed earlier, a recent US poll revealed that only two in five Americans describe themselves as being near the top of Maslow's pyramid, in the positive state we might call flourishing.[5] But Maslow probably never envisioned how many people in affluent, privileged positions are still stuck on the lower levels. Disturbingly, one in five—that's 60 million people in the United States—described themselves as experiencing the opposite of flourishing—languishing—with younger adults the most likely to be in this state of listlessness and unrest.

Considering how far we have come as a species, why are so many of us languishing? Obviously, our world still has plenty of problems—unstable autocrats, emerging viruses, economic and racial inequality, floods and famines, terrorist threats, and military aggression. Even some of the world's most privileged and developed nations seem to be coming apart at the seams. But still, for most people, in most developed countries, this is a pretty good time to be alive. Over the past century, advances in medicine and public health mean that infant mortality is way down and average life expectancy is way up compared to any other time in human history, pandemic-related dips notwithstanding. We can travel almost anywhere, associate with people from other cultures and religions, and—parental objections aside—date and marry according to the dictates of our own heart. Never have so many been able to meet their basic physical needs, the "ground floor" of Maslow's hierarchy.

Is languishing, then, to use a current pejorative, a "first world problem," meaning something that troubles privileged people who don't have any *real* problems? Are all those who feel stuck, frustrated, blocked, or near the breaking point simply ingrates who don't know how good they've got it?

We would suggest not. To get to the bottom of what's really going on, we can turn to the work of another luminary from the 1940s, Hungarian-Canadian endocrinologist Hans Selye, who introduced the world to the "general adaptation syndrome," more commonly known as the stress response. Selye described how we have evolved to survive when we become uncertain whether our resources are sufficient to meet life's demands. While some ambiguity around our capacity to meet a challenge can be exciting, too much challenge can quickly spiral out of control. Imagine the feeling of having to run one more lap when you are tired at the end of the workout, and compare that feeling to being told you still have ten more laps to go. The burst of excitement and determination at pushing for the finish line can quickly turn to worry and pessimism when the finish line seems impossibly far away.

THE BODY KEEPS SCORE

Selye's contribution lay in helping us understand how and why the breakdown occurs. Over decades of research, he demonstrated that our bodies respond to the presence of stress in three predictable phases: an initial alarm phase, where we become aware of the stress and are motivated to act; a resistance or adaptation phase, in which we attempt to live with the presence of the stressor; and, if the stress doesn't resolve, a final phase of exhaustion that's biologically catastrophic, sometimes leading to the death of the organism.

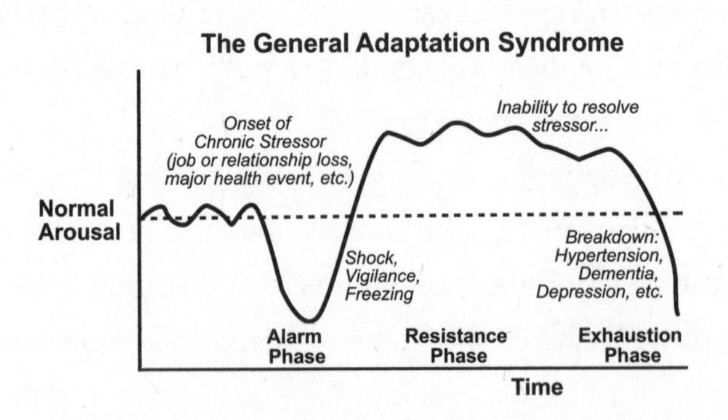

What makes Selye's work so relevant to our discussion is that, while technology and culture have changed dramatically since the Industrial Revolution, and the rate of change has skyrocketed since the Information Revolution, the hardware of our bodies hasn't changed much in at least a hundred thousand years. What we call "languishing" is consistent with hardware that's been pushed to the breaking point by unresolved stress.

But why are so many well-fed, safe, and reasonably secure people feeling such debilitating, chronic stress? For one, twenty-first-century capitalism is not for the faint of heart. Today's world seems to leave *everything* up to the individual, from health to career to retirement planning to twenty-seven different options for telephone service, all while we're expected to raise kids and perhaps care for aging parents, not to mention broader concerns about inflation, global warming, assaults on democracy, and a work culture that now expects you to be on call 24/7. So, the ambient stress level is often amped up to 11. Even when there are no immediate threats to life or limb, a lack of job security, past threats to our safety, or anxiety about the future nevertheless force us to watch for perils lurking just around the corner.

Persistent anxiety about an ambiguous future is exacerbated by our unparalleled media access to everything that's wrong with the world. It's statistically inevitable that, at any given moment, something terrible is happening to at least some of the billions of people with whom we're now electronically linked. We may be physically protected from that school shooting, or civil war, or tornado, but we feel the effects coursing through our neuroendocrine system, thanks to social media and 24/7 news.

And it's not as if the networks and platforms that link us to so many other lives represent reality in a neutral fashion. Twitter and

Facebook have come under fire for business models that seek to keep us engaged with shock and outrage and threats to our sense of safety and security. Meanwhile, they purposely promote content that undermines our sense of adequacy to sell us things we may not need.[6] Leaked internal documents show that when Instagram became aware of how negatively their site was affecting the mental health of teenage girls, their only response was to try to determine how to get these impressionable young women to spend *more* time on the platform.[7] There appeared to be little if any concern about the unrealistic imagery that the attention-capturing algorithms were masterfully pushing into receptive minds.

Our mental space has long been fought over on a battleground characterized by political scientist Herbert A. Simon as "the attention economy." And what is the most surefire way to capture someone's attention? To suggest an imminent threat. If we delve back into Selye's work, we'll see that our neuroendocrine system responds to the perception of danger with heightened vigilance and monitoring, which is to say, focused attention. If the threat signal can be sustained by incessant reports of outrage, contagion, or political extremism antithetical to your own views, your vigilance ramps up even higher. Each tweet or blast on your news feed hits your nervous system the way the sound of a breaking twig might create panic if you're walking home at night and you think someone may be following you.

We can't update our neural hardware, which evolved to manage hunting and gathering hundreds of thousands of years ago, but we can update the software. And we can do this by revising the "internal programming" we've developed to organize our experience and conduct our daily lives. This does not mean "improving" our habits. It means getting *beyond* habit.

MEANWHILE, BACK IN THE IVORY TOWER

Since the 1970s, physicians, psychologists, and others have offered techniques such as meditation and yoga for dealing with relentless stress, burnout, disengagement, and depression. While some approaches emphasize turning *off* the mind—a process of subtraction—we've discovered the powerful effect of addition. It was a paradigm shift for Western psychology, which had largely treated problematic thoughts as an infestation to be confronted and exterminated. The emerging mindfulness movement flipped the script, suggesting the problem was not the presence of such thoughts but treating these thoughts as special. Mindfulness suggested that a negative thought was just one of many unremarkable mental passengers, depriving a hurtful story of its mental gravity. Yet such techniques had largely eluded scientific description, shrouded in the Western mystification of Eastern traditions.

As a psychologist, coauthor Zindel Segal helped develop mindfulness-based cognitive therapy, among the most widely used training programs for promoting mental health. In the spring of 2004, he met coauthor Norm Farb, a grad student working on some of the first neuroimaging studies on how mindfulness training changes the brain. Together, we found that we had the perfect matchup of skills for leveraging mindfulness training to understand the transition from languishing to flourishing in the brain.

The two of us shared a strong interest in how new learning enables the brain to reorganize itself—a concept called neuroplasticity. At that time, psychotherapy, exercise, meditation, and yoga were all touted as having neuroplastic effects, but there was very little actual evidence. We set out to change that.

Work by Sara Lazar, Richard Davidson, and other scientific pioneers suggested that a lifetime of meditation could slow brain aging and synchronize brain waves. But we wanted to go beyond studies of those lucky enough to have undergone extensive training, to focus on the "thin edge of the wedge": how meditation helps everyday people better manage their emotions.

A clue from the clinical literature suggested that individuals who take things personally—the folks who view the remarks, gestures, and actions of others as being directed at themselves—often had greater psychological difficulties. So, perhaps meditation worked by reducing self-reference. After all, classic meditation texts hold out the promise of "non-self," "no-self," or, to quote musician and singer-songwriter George Harrison, dropping the "I me mine." Buddhism, one of the ancient traditions that gave rise to mindfulness and meditation, teaches that the self is nothing more than a mental fabrication, and seeing things through the lens of the self can be a source of suffering and division. From a Buddhist perspective, flourishing begins with holding a "beginner's mind," where you let go of expectations related to your self-concept. But how to measure a person's degree of self-centeredness? Fortunately, we were able to tap into a line of Western research three decades in the making.

In the early 1970s, Fergus Craik and colleagues at the University of Toronto demonstrated the role of self-reference in how well we retain information.[8] They showed that study subjects who were asked if a word applied to them as individuals remembered the word better than subjects who were asked something less personal, such as whether the word was printed in upper or lower case. The discovery that the brain prioritizes self-reflection above other types of thinking was an important finding, validating the

Buddhist characterization of mental life but also helping to explain why people get stuck in negative self-evaluation. If self-judgment is a prioritized mental process, then we can begin to understand Selye's description of becoming locked in a resistance mode leading to burnout: resistance begins with a self-judgment of "I'm not sure I can handle this," and the brain prioritizes these judgments to the exclusion of other options.

In 2005, we ran our first neuroimaging study with a similar protocol.[9] We asked participants to *judge* whether certain words described them or not, a lab-controlled version of "am I good enough" judgments that trigger the stress response. Previous studies had identified the brain regions supporting self-reference as being along the brain's midline, more specifically in the medial prefrontal cortex.[10] It was not terribly surprising, then, that the midline of the brain lit up for our participants who'd been asked to judge.

Then came the critical test. We asked participants to read the trait words without any self-evaluation. Instead, we asked them to *sense* any physical sensations, feelings, or thoughts they might have in response to these words. We hypothesized that *sensing* would suppress the conceptual, reasoned self, and the impact of self-judgment along with it. After all, if you get rid of the self, there's no one to have low self-esteem, guilt, and hopelessness, right?

We knew that the deck was stacked against our participants' ability to shift to sensing. The memory studies from the 1970s showed that the brain prioritizes self-judgment over nearly any other type of thought. More recent work dubbed the brain structures imposing this priority as the default mode network.[11] The DMN dominates mental life even when the brain is at rest, diverting resources away from sensory parts of the brain.

To give ourselves a fighting chance of demonstrating an escape from this strong bias toward self-reference, we enlisted people who'd signed up for an eight-week course to learn how to "transcend the ego" through meditation. We randomly assigned half the people to be scanned using magnetic resonance imaging (MRI) after completing the course. The other half had signed up for the training but had not yet received it, what is known in the business as a "waitlist control."

By 2006, we'd completed one of the world's first neuroimaging studies of mindfulness training. We were getting good data showing self-evaluation lighting up the DMN. We'd validated our mindfulness training program, developed state-of-the-art methods for brain imaging, and put together a great research team with a wealth of both clinical and neuroscience experience. We even validated the training, because the people who'd completed it were feeling much better than the control group who'd been wait-listed to take the course but had not yet participated. The big question, though: Would we be able to show that meditation reduced stress by breaking down the DMN's relentless push toward self-evaluation?

We pored over the functional MRI (fMRI) data, looking for specific evidence that meditators were able to turn down the tendency to view themselves judgmentally. Our hypothesis showed limited success: when people focused on sensing their experience, we saw less activity in DMN brain areas that supported judgment in both the untrained and trained groups alike.

However, we couldn't find much evidence that people trained in mindfulness were *better* at reducing self-judgment. We tried every trick in the book to illuminate this effect and spent nearly a year redoing analyses. We sat in meeting after meeting talking about how this "self region" in the brain just wouldn't turn off, bemoaning the fact that mindfulness training wasn't supporting our theory. We kept brushing past the fact that we had evidence of training effects in other parts of the brain. That's because they weren't where we expected them to be. We weren't seeing what we were looking for, namely, the curtailing of conceptual self-reference. Instead, we *were* seeing greater activity in seemingly irrelevant parts of the brain that supported body sensation.

And then it hit us. Mindfulness meditation's benefits don't come from *getting rid* of the self. What the data showed was an *expansion* of the self in the form of immediate sensory input. The negative self-judgments are weakened not by "judging judgment" and suppressing the impulse but by enriching the self-concept with novel, conflicting, disruptive, and ultimately creative experiences. Rather than turning *off* the parts of the brain that support judging, people trained in sensing had learned to turn *on* other brain regions dedicated to sensation. Bottom line: despite starting off just as stressed, anxious, and depressed as the control group, the mindfulness group felt better after training.

We all know intuitively that a blast of sensation can lift our mood. Stepping away from thinking and striving, stepping into our body's ability to sense the world—watching waves crashing on a beach at sunset, smelling lilacs—can momentarily ease stress and clear the mind.

What we don't appreciate is how this natural, restorative response has in each of us been suppressed by a lifetime of counterprogramming. We are told as children to "stop dawdling and focus"; as teenagers to "take responsibility and focus on your schoolwork"; as adults to "put in your time and grind until you are successful." Even if each of these admonishments is well-intentioned, together they carry the consistent message: sensation is a waste of time unless it helps you do something useful. When we're growing up, nobody tells us that sensation is a capacity that withers in the face of stress. If that capacity is left neglected, we may be putting our mental health at risk. And after a lifetime of such neglect, it is far less likely that we will avail ourselves of the benefits of sensory awareness, especially when we need them most.

Exercise: Meet Your Senses

Immersing yourself in sensation might mean watching the sun set over the Pacific from a cliff near Big Sur or taking in the salt and ozone smell of a tropical rainstorm from under a shuddering palm tree. But nothing says it has to be that majestic.

The participants in our studies learned to engage in sensation right where they were, as part of their daily routine. It takes practice, though, partly because our senses are so adept that we usually take their work for granted.

Here are some exercises to get you started. You can do them anywhere, but it is important that you set time aside and eliminate distractions. You get what you pay for, and in doing these exercises you really need to pay attention.

1. **Seeing**

   Your eyes translate light into complex image data for the brain to decode. Seeing is considered the most powerful of the senses because we rely more on sight, compared to hearing or smell, for collecting information from the environment.

The practice: *Whether you're indoors or outdoors, look around you and pick out five things that you can see, and name them or write them down. Then close your eyes for a moment. When you open your eyes, look around you again, only this time, pick out five things you didn't see the first time, maybe because they were in the background or were lost in a busy scene. Take a moment and actually name each thing that you notice.*

2. **Touching**

Touch, perhaps the very first sense that humans develop, works through specialized receptors just under the skin to convey signals about pressure and temperature to the brain. Being the body's biggest organ, the skin can provide a quick readout on what you're experiencing.

The practice: *Whether you're indoors or outdoors, ask a friend to collect a variety of small objects for you and place them in a bag, so that you don't know what to expect. These can include anything: a rock, a coin, a pencil, a tennis ball, shells, leaves, rice. Then close your eyes and place your hand or hands in the bag and feel each item, exploring its contours and dimensions. You may be able to recognize and name the object, but more important is to notice what it is like to touch without knowing, to see what shows up before the label is applied.*

3. **Hearing**

Hearing results from sound waves in the air hitting our eardrums, which transfer vibrations through bones and fluids to send sound signals to the brain. Sound waves vibrate at a unique frequency, and when they arrive at the ear, they are amplified so that the brain can distinguish between speech, laughter, music, and the rest of the auditory experience.

The practice: *Search your memory for the last few times you walked outside, then make a list of sounds that you might have heard. Depending on where you're located, these might include a car approaching or honking, birdsong, people talking, a door slamming, wind moving branches, subway doors closing, a dog barking, footsteps on grass or gravel. Now head out and listen for these sounds, and as you hear each one on your list, cross it off.*

4. **Smelling**

Smelling starts when we inhale through the nose. Chemicals in the air hit cells on the roof of the nasal cavity and olfactory cleft that transmit information to the brain to decode as smells. It is estimated that we can detect more than 1 trillion different scents.

The practice: *Find five things in your house or outside, some familiar and some unfamiliar, and smell them. As you inhale, try to label each scent as vividly as you can, naming it if possible but also considering what you most associate it with. For example, the citrus tang of an orange peel might call to mind clementines in wooden crates, eaten with family during the holidays. Some smell starting points might be coffee grounds, a bath towel, cooking spices like cinnamon, ginger, or curry powder, a bar of soap, boiling water, potting soil, your underarm, wet leaves. Smell and memory are located close to each other in the brain, which explains why smells often trigger recollections. Pay attention to see if any of these smells produce this effect.*

5. **Tasting**

   The tongue is our principal taste organ. Thousands of taste buds break down tastes into categories of salty, sweet, sour, bitter, and savory. Chewing and swallowing lead to tastes from different parts of a meal combining in our mouths. Each of these tastes is sensed by the tongue, and these taste data are sent to the brain. The brain then decodes this information to provide us with flavors we can identify.

   The practice: *The next time you eat something, see if you can name each distinctive taste—salty, sweet, sour, bitter, or savory. This can be during a snack where you might be eating only one thing, or during a meal where you will be eating a variety of foods. When you recognize one of the principal tastes, say its name out loud.*

Like our study participants, you may have noticed that it isn't always easy to immerse yourself in pure sensation. Often, there are other things going on in our minds, even when we try to focus and pay close attention. The amount of effort it takes to remain focused on sensation underscores the stickiness of more familiar ways of relating to our experience, the habits and mental routines that structure our mind.

2

THE HOUSE OF HABIT

We are what we repeatedly do.

—WILL DURANT

LIFE IN THE DEFAULT MODE NETWORK

Since toddlerhood, we've been conditioned to pay attention, to solve the problem, to not be distracted, to do our best, and above all, to defer gratification until the task has been successfully completed. The result is a wide array of mental and emotional challenges that come from spending too much time thinking—especially about how we're measuring up—and not enough time exploring for exploration's sake, enjoying the sensory world around us.

Being scientists does not grant us immunity from these tendencies. In our first study, we were so invested in tweaking judgment that it took us a long time to see what the data were saying. A massive blind spot in our initial analysis kept us from seeing that there was something more important than higher cognition driving well-being. Somewhat ironically, one reason for our failure was the network responsible for our single-mindedness: the default mode network. Every time we looked at the data expecting to see the DMN deactivating with mindfulness training, every time we felt surprised and disappointed when we failed to see movement in DMN activity, it was our own DMN that was getting in the way of fresh insight.

The House of Habit formed by the DMN extends like a dorsal fin atop your brain, with two additional hubs, one above each ear. It is not only the epicenter of pure "self" consciousness and the center of judgment-making but also the part that habituates us to focus on what *it* sees as the most essential tasks of survival. The DMN matches learned responses (habits) to the complexity of the specific situation before us at any moment. This conservative impulse to keep us in the realm of habit is an enormous advantage when we need to be efficient with our attention. It's the DMN that takes the effort out of repetitive, automatic acts (sometimes referred to as "mindless activities") such as brushing our teeth. The DMN also creates templates for more complex but still routine behaviors, such as driving to work or giving a well-rehearsed presentation. It allows us to navigate familiar places, move through human or vehicular traffic, or prepare familiar recipes while still having the mental resources available to consider the future or take in music or a podcast. More broadly, its readily available supply of off-the-shelf responses allows us to maintain a sense of who we are and how we fit into a larger group. Without the DMN, we'd have to relearn this contextual information each day, which would leave little time for much else.

But there's a cost to maintaining a full range of automatic responses at our fingertips: the need to quickly match situation to

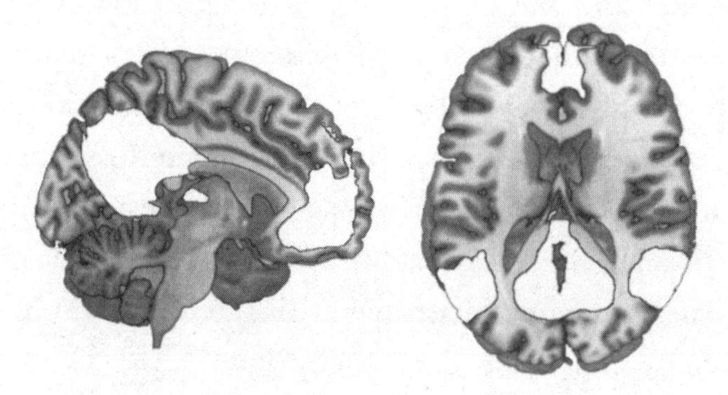

The default mode network. Left panel: side view; right panel: top-down view.

habit to ensure survival often keeps a "resting" brain as busy as a striving brain. Given that our default is to try to solve problems, to compare ourselves to others, to ruminate over the past, and to plan for what the brain anticipates as being the future, the closest we come to mental rest is usually through mind wandering. But even when the mind feels free enough to wander, it doesn't do so randomly. Instead, it follows well-worn paths that can become counterproductive, exhausting, and debilitating ruts.

Perhaps the most impressive demonstration of the DMN's centrality in human life is a series of studies that show that allowing a person to disengage from focal tasks in the scanner is one of the most reliable ways of activating the DMN.[1] At the end of some effortful task, such as memorization or mental rotation, we often tell participants to "rest and relax." The moment they "let go of the wheel" and stop directing the mind, the DMN leaps into action, spinning up narratives dealing with plans, regrets, obligations, and desires. It's been suggested that the DMN is the biological seat of the stream of consciousness, or the internal monologue that we all experience *as* consciousness.[2]

The naming of the DMN only took place right around the turn of the new millennium, but it was more than a century ago that Hans Berger, the inventor of the electroencephalogram,[3] showed that electrical activity in the brain remains high even when the mind is supposedly at rest. (Perhaps this is what the French mathematician Blaise Pascal had in mind in the mid-seventeenth century when he famously attributed all the world's problems to man's inability to sit alone in a room by himself.[4])

As neuroscience technology evolved in the 1950s, we gained the ability to measure metabolic activity in the brain—how much oxygen and sugar are consumed at each moment—with an imaging technique called positron-emission tomography (PET). Making use of PET scans, Louis Sokoloff and his team made the surprising discovery that moving from rest to doing math problems did *not* alter brain activity.[5] In the 1970s, David Ingvar noticed that some frontal parts of the brain were actually *more* active during rest than during "effortful" mental processing.[6] Studies like these gave rise to a theory that perhaps all the internal monologuing we do simply doesn't register as effort because it's so pervasive and persistent as to seem natural. It's just something you do, like taking your next breath.

Exercise: Open House at the House of Habit

While at first blush it might seem extraordinary that there is a brain system that's always expending energy, never really resting, you can test this idea for yourself. After you finish reading this paragraph, stop reading for about thirty seconds and notice what happens inside your mind. No need to do anything special, just let your mind do what it wants to do naturally, without having to focus on anything or get the "right answer." Ready? Go for it!

. . .

. . .

. . .

And . . . we're back! We will now make three predictions, based on our years of psychology training and a healthy dose of chutzpah. The first is that *your mind wasn't very quiet or restful at all.* Chances are you wondered something along the lines of: "What am I supposed to be doing? Am I doing it right? I don't sense anything! This is stupid!" Or maybe you felt a twinge of anxiety as you tried to count down thirty seconds. Maybe you made a judgment that it's silly to follow an exercise in a book, or worried that the experience wouldn't live up to expectations. Maybe you felt excited and curious about the chance to explore. Maybe your mind went immediately to "What's for dinner?" But we can safely assume that you didn't go completely blank for thirty seconds then start reading again.

Here's prediction two. Whether you experienced a cacophony of self-talk or just a few clear moments, we expect that a lot of your thoughts weren't about art or justice or algebra. Most likely your thoughts were about you! "How long am I going to do this exercise? What am I going to get out of this? What am I supposed to be feeling?" I, I, I, me, me, me! And why not?

For our third prediction, let's guess that when you let your mind "rest," not only did you experience automatic thoughts, and not only were these thoughts about yourself, but they were also very, very *familiar*. By this, we mean if you and three of your friends each transcribed your thoughts over the thirty-second period, you could likely pick out whose thoughts belonged to whom. If you're someone who gets bored easily, you may have thought about being bored and time being wasted. If you're overworked, maybe judgments of fatigue showed up. If you're a foodie, perhaps dinner was on your mind. If school or work is big in your life, maybe that evaluation or deadline crept in. And if you're experiencing relationship drama—guess who showed up in your quiet moments?

While a few sentences of internal monologue in a moment of mind wandering may seem trivial, we suggest that it's anything but. This is the DMN drawing you back into the habitual, perhaps even obsessive, patterns of your life. Despite its many negatives, the DMN's self-absorption can serve you well as a counterweight to becoming overly caught up in an activity or someone else's agenda. It is the little voice that asks when you can take a break during an arduous activity or questions the relevance of a boring lecture. While a lot of the brain is set up to solve problems and engage with the world, the DMN reminds us to take care of ourselves along the way, reinforcing a selfish context to make sure we don't forget about the bigger picture. You may have gone out and solved a math problem, planned your travel route, or been immersed in a moment of conversation. But when all is said and done, the DMN brings you back home to the House of Habit.

THE BENEFITS OF HABIT

Having the DMN insistently spin up self-reflective thoughts and habits can be exhausting, but as noted, we'd never make it through the day if we had to effortfully plan out who we are and what we're doing in each moment. There's simply too much to figure out and never enough time, which is why much of life needs to be conducted on automatic pilot.

Nobel laureate Daniel Kahneman's *Thinking, Fast and Slow* brought widespread attention to automated behavior and its opposite. The book provides a detailed account of how the brain alternates between slow, deliberative reasoning (System 2) and a rapid, intuitive state (System 1). Neuroscience backs up this observation. It also shows how the conscious, deliberative work of the "central executive" parts of the brain (System 2) is turned over to the DMN (System 1) after a bit of repetition. Our own neuroimaging work shows that the DMN remains quiet when participants first engage in repetitive activities, like pressing one of two buttons to reflect the direction of an arrow on a screen (⬅ or ➡). Within five to ten seconds, though, as the person settles into the task, the planning and attentional control parts of the brain quiet down and the DMN activates. For more complex behaviors, such as learning to play a musical instrument, it takes far longer to get beyond intentional movements and let "muscle memory" take over. In his book *Blink*, Malcolm Gladwell suggests that ten thousand hours is the usual benchmark for mastery. In neuroscience terms, Gladwell is describing how the DMN slowly integrates newly acquired skills into accessible and seamless habits for perception and performance.

For example, consider a skill that most of us have mastered: standing up. Imagine what it would be like if every time we wanted to stand, we had to think, "Okay, lean forward, engage leg muscles, push against the floor, start breathing slightly faster, speed up heart rate, fixate on a reference point, compare visual motion against the swish of liquid in the inner ear to find balance, send signals to the muscles to adjust, et cetera." Instead, all these coordinated actions take place instantly and effortlessly. If you watch a toddler learning to stand, you can appreciate how this encoding takes hundreds of repetitions before it becomes fluid. Conversely, we can see how these habits break down as we age or if we suffer an injury.

While early life skills such as walking, swallowing, and breathing are eventually automated via the brain stem and other, more "primitive" parts of our brain, the DMN manages more complex processes, connecting incoming sensations to our social context, mood, and motivation. The DMN also helps us attach emotional import and significance to our experience.

In a world where we must constantly make decisions amid seemingly endless cascades of information, having a way to automatically reduce the chaos and complexity of each moment into a consistent, sensible reality is a great advantage. The philosopher Daniel Dennett describes this reduction as being a fundamental aspect of being human, a type of "narrative gravity" that gives us a coherent sense of self. Not surprisingly, Alzheimer's disease targets the DMN first. Even if we sometimes pine for the simplicity of childhood, the involuntary descent into a more childlike state is no picnic. As growing numbers of families now know, the disorientation and dysfunction that follow becoming untethered from context create crushing disability.

Just as we take for granted that our hearts need to beat faster *automatically* when we start moving, we take for granted that we have automatic access to the conceptual information we need to navigate the world each day, and it's here that the DMN really shines.

Having a relatively universal brain system for maintaining our sense of identity and personality also allows us to expect the same consistency from other people. The idea that knowledge is automatically encoded and available to us is the foundation for human integrity and commitment. How could you hold someone accountable for making a promise if knowledge of that promise wasn't assumed to be automatically carried forward into the future?

This shared brain system also makes daily social interactions run much more smoothly. Imagine that you run into a friend on the street and they ask how you're doing. Thanks to the DMN, the question doesn't inspire rigorous self-examination. We don't stand there with a puzzled expression, wondering, "Well, how *am* I?" Instead, we have a readily available sense of how we're doing, as well as practiced habits for how to respond. The responses are automatically modulated by whether this is a person we should be casual or formal with, and by whether we should be entirely candid or simply give the conventional response "Fine, how are you?"

We go through this routine so automatically that it's disturbing whenever the process goes awry—when someone "overshares" some intimate detail in a formal situation, or, conversely, when a romantic partner seems unable to get beyond surface-level interactions even after months of dating.

As human beings, we have a deep, fundamental motivation to make sense of the world and to be accurate in the picture we paint. One way we get there is by having background models, or "heuristics," for perception and action, and it feels great when they run like clockwork. You can think of the DMN as a concierge for such heuristics, the ultimate explainer of life experience. So, the DMN is doing more than just providing an inner monologue and a sense of self. It helps us automatically understand that a red light means stop, that we should hurry because we're late, that a person moving toward us is a friend and should be greeted differently from a stranger.

WHEN AUTOMATION GOES WRONG: THE SORCERER'S APPRENTICE

Given all the benefits the DMN provides by maintaining the House of Habit, how does it turn into the DAMN network, leading to frustration, resignation, or burnout?

The answer is that, just because information has been encoded and automatically accessed doesn't mean that it applies to the situation at hand. And when the well-established knowledge we're exploiting is incorrect or inappropriate, it can lead to serious trouble.

If you visit the United Kingdom, Australia, or Singapore (or you're from one of these places and visit almost anywhere else), you must continually check yourself to avoid getting run over, because everyone is driving on the "wrong side" of the road. That's why the 30 percent of countries (mostly former British colonies) whose citizens drive on the left side of the road have giant painted wording in the crosswalks warning "LOOK RIGHT." They know that an intellectual understanding of the local custom is no protection when up against deeply ingrained habits.

Part of our automatic, "System 1" understanding of the world is which direction cars come from. This understanding is so baked in that it requires effort every time you approach a road where this rule does not apply to overcome existing knowledge. You're *stuck* with this knowledge, at least for the first few days, weeks, or months you spend in a new place where your knowledge obviously doesn't apply.

But not all contrasts between the habits of existing knowledge and new information are quite so stark. We can be aware that something is wrong but feel powerless to do anything about it. We simply watch ourselves mess things up, yet again. This is when the DMN can be truly pernicious. Confronted with disruptive new information, the DMN doubles down on the old patterns of behavior, insisting that we do the habitual thing the habitual way.

If you've seen the 1940 Disney film *Fantasia*, you'll remember Mickey Mouse as the sorcerer's apprentice. One of Mickey's daily chores is to bring water from an outside well and fill a stone tub in the house. An enterprising mouse, he's always looking for ways to make his work life easier. He watches intently as his master, Yen Sid, wearing his sorcerer's hat, conjures spirits over an open fire. When Yen Sid grows weary, leaves to get some rest, and leaves his magic hat behind, Mickey sees his opportunity. He places the hat on his head and, imitating Yen Sid's gestures, uses the hat's magical powers to automate his chores. He commands the broom to fetch the water, which works so well that, once the broom seems to get the hang of it, Mickey decides to take his own little siesta. When he wakes up, he finds the room filling with water and the broom maniacally adding more bucketsful to the already brimming tub. To forestall a flood, Mickey grabs an axe and chops the broom to pieces, but then the shattered splinters come to life, each one now carrying in even more buckets of water. Only Yen Sid, once he returns, is able to use his magic to make the broom splinters stop.

The lesson for Mickey, and for us, is that automated behaviors are helpful only when they fit the situation. It's great not to have to think about operating an elevator or pouring soup out of a can, but what happens when we automate more complex tasks, such as reading emotions in a social situation, or addressing hurt feelings after an unkind word? We may have gotten so used to laughing and shrugging at a friend's wry sarcasm that we laugh and shrug when she's actually upset. Any time we follow the DMN's lead for a quick take on the world, we run the risk that we'll miss the nuance and match the wrong behavior to a seemingly similar but actually quite distinct template.

THIRSTY ZEBRAS

By providing patterns for us to follow—"rules of thumb" for every situation—the DMN, in essence, fosters a system of *knowledge exploitation.* It consolidates what we know and tries to trigger the appropriate set of instructions for managing the incredible complexity of everyday life, and with the greatest efficiency. But being locked into a pattern of exploiting existing knowledge, to the point that we can no longer consider alternatives, is also called being "stuck." Like a zebra returning to the same dried-up watering hole after it has been depleted by drought, we suffer when our own *default mode* of exploitation persists in the face of a changing reality.

Shanice was like that thirsty zebra when she kept working the way she'd always worked until she was totally burned out. We are often no different: whether with an employer, romantic interest, or family member, many of us return to harmful relationships for fear of being unvalued or alone, and because toughing it out seems safer and easier than starting all over again. In hindsight, such failures to change don't make much sense, but we don't consciously decide, "I'll just go ahead and stay in this miserable job/relationship/town/apartment even though it's ruining my life!" It can take quite a while before we realize that something just isn't working anymore, and even then, without a clear alternative—a new opportunity waiting in the wings—it just never seems like the right time to make a change.

In *Fantasia*, it's obvious that Mickey's attempt to automate his chores is going to get him in trouble, because he's playing with forces he doesn't understand. Yet as much as we may feel superior to Mickey in his dilemma, the DMN provides each of us with a similar set of magic brooms that can lead us to trouble. What are your "brooms" doing? Do any of the following sound familiar?

- I do my best not to argue with my partner but still find myself embroiled in daily conflict.
- I promise to be more open but mumble "I'm fine" when I'm really upset.
- I organize a to-do list to be more efficient at work, then spend hours on social media.
- I promise to stop procrastinating and still find myself hastily cobbling together materials moments before my presentation.
- I succeed in not snacking at 9:00 p.m. but find myself covered in crumbs by 9:30.

WHEN AUTOMATIC KNOWLEDGE WORKS AGAINST MENTAL HEALTH

Given that the DMN handles all things automatic, it plays a significant role in how we react to sad moods. Unfortunately, in this case, its go-to strategy of reverting to habit often means relentlessly judging the self.

If someone with a history of depression asks, "Why do I feel bad right now?" their DMN will very likely supply an immediate set of follow-up questions: "What have I done wrong? Whom have I failed to please?" Someone who's never had a problem with depression, though, will just as likely wonder if their blood sugar is low, or shrug and say, "Everyone has a bad day now and then."

It's not unusual to stew occasionally when feeling down. But if your habit is to see a negative mood as a sign of how useless and hopeless you are, the DMN runs with it, recommitting to your expectation of anger and rejection from the social world, which limits your ability to notice any pathway out of your funk. When feeling bad is all too familiar, you begin to forget that you are not condemned to always feel this way. When a depressing life becomes the unsurprising norm, it saps the motivation to look for an explanation other than "it's all my fault." That's because the DMN demands an answer—any answer—as it attempts to make sense of what's going on, even if that answer is, well . . . depressing. Unfortunately, this quick-and-dirty approach robs us of the hope that feelings might change over time, or that our mood might be more nuanced than it initially appears. The prepackaged context supplied by the DMN locks into place our initial, self-blaming impression of a negative emotion, which then justifies and explains the feeling according to the DMN's readily available rule of thumb.

The DMN's mental routines evolved to help us survive long enough to reproduce, but they are agnostic when it comes to our individual well-being. Cues that supply mood and context, for example, can improve recall of related themes, which is a good thing for a singer trying to remember lyrics. If she suffers from PTSD, however, and wants to avoid triggering a traumatic memory, those same cues can be devastating. Our frustration at not being able to shake off the negatives may lead to anger, but that merely tightens the DMN's grip.

Well over a century ago, Sigmund Freud described how quickly memories of an emotional event can be triggered by similar situations to form a mental representation, or complex, that keeps these ideas and emotions alive in a person's mind. For someone with what Freud called a persecution complex, even a minor disagreement can trigger the feeling that people are out to get them. Someone with a god complex will respond to appeals to follow social norms with well-rehearsed reasons for ignoring them. Freud collected colorful case reports (the evidence for "complexes" is largely anecdotal) that illustrated how these tendencies interfered with interpersonal functioning, but his narratives say nothing about the mechanisms that cause the interference. While Freud's theories have given way to more systematic forms of research, the question remains: are there hidden mental habits for perceiving and responding that can help us understand why we sometimes have trouble coping? Thinking back to the DMN's role as a "concierge of heuristics," meeting every experience with ready-made explanations and expectations, we can see that it represents a strong candidate for what Freud was talking about. Could the DMN inadvertently be keeping rigid thinking styles in place?

We tackled this question by studying people likely to have such "complexes" who had recovered from one or more episodes of depression. These individuals appeared relatively healthy and were hoping to stay that way, but their clinical histories suggested a 40–50 percent chance of becoming depressed again, trapped in the DMN's habitual reinforcement of their negative worldviews.

Clinical research confirms that depression is more likely when you've been depressed before, which supports the idea that some internal programming has gone awry. Understanding the DMN-fueled propensity to repeat maladaptive tendencies helps to explain why depression and anxiety often become chronic and recurrent disorders. To find a way to help prevent depression's return, we needed to understand the role of the DMN's automatic emotional processing, specifically, the kind of processing cued by negative moods.

In the early 2000s, just as Segal and Farb were getting involved with brain research, Segal helped set the standard for studying psychological vulnerability for depression. In one of his groundbreaking studies, he canvassed hundreds of individuals receiving treatment for depression and ended up with ninety-nine participants who were far enough along the road to recovery to enter a follow-up program.[7] To see a depressive context playing out in real time, he invited these people into the lab to listen to a piece of music likely to induce a sad mood. This was Sergey Prokofiev's orchestral introduction to the film *Alexander Nevsky*, a tune with the snappy title "Russia under the Mongolian Yoke." Trust us, if any piece of music can be a downer, it is this one, but just to be sure, the team remastered it to half speed. At the same time, they asked these study subjects to recall an incident in their lives that had made them feel sad. This combination of lugubrious music and sad recollections induced a mild sad mood that lasted between five and ten minutes, then cleared up.

Segal had asked participants to fill out a measure of depressive thinking styles before and after the experiment. A classic example is "black or white" thinking about personal adequacy and self-worth, thoughts such as "If I don't do well in everything, I'll never get respect" or "My happiness depends more on other people than it does on me." Everyone has occasional thoughts like this, but some people show a really exaggerated tendency to endorse these extreme assumptions when they're feeling down.

After the assessment, each person got a score that indicated their tendency to fall back into such harsh and absolutist patterns of thought. The team then followed these folks for eighteen months to see who relapsed and who stayed well. What was fascinating was that participants' responses to sad moods in this simple experiment predicted how they would fare months later. Comparing their sad-mood to normal-mood responses, 70 percent of participants seemed *nonreactive:* they were unaffected by the mood induction, showing either no change or a decrease in dysfunctional thinking. The remaining 30 percent were *reactive:* when placed in a sad mood, they endorsed a significantly greater number of dysfunctional thoughts, indicative of more "black or white" thinking.

Strikingly, this cognitive reactivity to a sad mood had important implications for how people fared during the follow-up. Only 30 percent of the nonreactive group suffered a relapse. On the other hand, 69 percent of the reactive participants suffered a clinical relapse—indicating a more than doubling of the risk. This outcome showed, for the first time, the hidden fragility that can persist after a person climbs out of depression. At the start of the study, everyone had recovered from a depressive episode and had normal scores on clinical measures—they would each tell you honestly that they were doing well. The vulnerability, however, was hidden below the surface, a melancholic predator waiting for signs of distress before striking its prey. Yet depressive thoughts and memories that went unnoticed when things were calm raced to the forefront at the first sign of stress.

When triggered, the context of past suffering so permeates the mind, and so skews habits of thought, that it can dominate an individual's experience of the present. Everyone, our participants included, inevitably encounters setbacks and challenges, but the vulnerable fall back on dysfunctional attitudes that impede adaptive coping. In the face of stress, it seems that the vulnerable are held captive in the House of Habit.

When we first started working together, we reflected on what might be happening in the brain to explain Segal's provocative findings. At this time (the mid-2000s), theories of brain networks were just becoming popular; to us, depression vulnerability seemed like the workings of the DMN. It didn't take a great stretch of the imagination. The DMN could just be doing what it always does, automating perception and response, but in this case, imposing some very rigid and unfortunate programming.

Around the same time, our intuition was supported by converging findings from an independent group of researchers who reported that DMN hyperactivity was common in depression. Led by Yvette Sheline at Washington University in St. Louis, the researchers had compared the brains of healthy people and depressed people while both were "at rest" in the scanner. Sheline found that in depressed folks the DMN had spread well beyond its normal boundaries, taking over brain regions that normally would be involved in planning and emotion regulation.[8]

This was an important finding, because DMN encroachment on areas for planning and decision-making represents a collapse of the critical distinction between being on automatic pilot and having control over one's life. In depression, even effortful and intentional thinking becomes biased toward questioning one's adequacy. Every issue boils down to "What does this say about me?" No matter how hard the depressed person tries to stay on an even, non-self-judging keel, he or she is like a long-suffering younger sibling, thrilled to be playing Nintendo with their older brother without realizing that his controller is unplugged. That's part of what is so insidious about mood disorders: it doesn't matter how much we twist the steering wheel, our cart continues along a fixed track.

For the person with depression or anxiety, that track consists of unhelpful rumination rather than true problem-solving. With Sheline's work, we had clear evidence that an overexcited DMN could predict an internal world frantically rehearsing pessimism and helplessness, one that shuts down efforts to go "off script" by constructively reinterpreting experience as an irrelevant distraction.

Exercise: Come On, Feet—Getting Moving

We imagine by now that you're intrigued but maybe skeptical about how much of your life is programmed by your DMN. For a simple, down-to-earth example, let's look at your feet.

These guys take a real beating—walking for miles, wearing uncomfortable shoes, freezing in snow, or stepping on stones. We sometimes try to compensate with pampering—polishing, soaking, and scrubbing. Usually, we pay attention to our feet only when they give us trouble. If that occurs, the mind snaps to attention. Here's a discrepancy that needs to be resolved—right now!

But take a minute or two to think about your feet *without* looking at them. What comes to mind? Shoe size? That's kind of an automatic, objective fact. But some subjective judgments probably spring up as well. Do you like or dislike your feet? Would you want them to be different? Are you worried about whether they smell? And what's the aesthetic status of your toenails?

Maybe you imagine the places your feet have taken you, or the problems they've caused you, or you judge them for having become ugly over time—your toes now a far cry from the "little piggies" your mother would have kissed. Just go wherever your mind takes you. The point is, you have a lot of stored knowledge about your feet, and all you have to do is think about them in order for all this information to spring up unbidden.

Now, for a twist, take a minute or two to pay attention to your feet *while* looking at them. Let your attention sink into your feet, inside and out, from the bones right out to the skin itself, feeling the sensations of touch on the skin, sensations of pressure where they contact the floor or bed. Now go ahead and clench your toes, curling them in as close as you can, paying attention to the physical sensations in the toes and in the soles of each foot: muscle tightness, the flow of sensation all the way up to the ankles. Then unclench your toes, noticing any changes in the sensation as your muscles relax.

When you compare *judging* to *sensing*, even a seemingly mundane target like your feet can make very different things pop into your mind. Judging seeks to establish certainty; there is a gauging of "What do I need to do here?" Get a pedicure, see a podiatrist, wear a larger shoe size? Is there a gap between how things should be and how they are? Sensing can feel less certain in that there's no fixed outcome, but it can also be less stressful because judgment recedes to the background. When sensing, did you notice that searching for a discrepancy was less of a priority? If you find that you can feel your toes and feet but have no idea where all this is going, congratulations, you're ready for the next chapter.

3

KNOWING IS HALF THE BATTLE

I believe . . . that the petal of a flower or a tiny worm on the path says far more, contains far more than all the books in the library. One cannot say very much with mere letters and words.

—HERMANN HESSE, *NARCISSUS AND GOLDMUND*

THINKING LIKE A MAMMAL: THE EXPLORE/EXPLOIT TRADE-OFF

Remember Shanice, the plucky entrepreneur facing burnout as she struggled to manage a growing business? Launching the business of her dreams locked her into a furious and ultimately unsustainable pace in which Shanice the individual disappeared into a blur of activities and obligations.

Through no fault of her own, she'd lost sight of a fundamental aspect of survival for most organisms—*the explore/exploit trade-off.* An ant colony might thrive on a rotting log, an abundant resource that's very efficient to exploit. But sooner or later, resources run out, so the colony devotes a portion of its energy to exploration, sending out scouts to *forage* for the next source of nutrition. Smart businesses do the same thing, always foraging for new lines of revenue. The alternative would be to repeat the dismal lessons learned by the recording industry, which was completely blindsided when streaming came along and revolutionized how people listened to music.

Shanice is only doing what comes "naturally" to her, and to the rest of us—exploiting existing knowledge in the face of uncertainty, fear, and threat. She can't see that just sticking with her "work harder" strategy has stopped yielding returns, like going back to the same food patch when its resources have been depleted. She's so trapped in the House of Habit, stuck trying to exploit the resource she knows well, that she doesn't realize another option might exist.

SENSATION IN THE BACK

Studies of the natural world tell us unequivocally that exploitation alone is not the best survival strategy. Natural selection favors foraging strategies that flexibly balance the benefits of exploiting an existing resource while continuing to keep an eye out for new possibilities.

Fortunately, we're built to strike this optimal balance. The DMN wires us to strap on the blinders and exploit the tried and true, but much of the rest of our mental apparatus is dedicated to exploring the landscape of possibilities. We need input from our senses to do this, and there are wide swaths of the brain dedicated purely to sensation.

In fact, the distinction between sensing and responding is one of the most fundamental ways the brain is organized. You may have heard of other ways of mapping out brain regions and functions, such as the "right brain is for feeling / left brain is for logic" dichotomy. In today's science of neural networks, claiming that any abstract concept is physically localized in one area seems . . . well, bogus.[1] By contrast, the building blocks of concrete, sensory experience are well delineated.

Hundreds of thousands of scientific studies now confirm the idea of distinct sense neighborhoods, as well as their specific locations. In these studies, vision is *always* represented at the very back of the skull, hearing just behind your ears, smell and taste at the top of the nose, and touch along the "hairband" of gray matter that runs from the crown of the head toward the ears. Even the sense of what's happening inside our bodies—interoception—has dedicated real estate, located just underneath that somatosensory hairband that descends in front of the ears. Interoception gives us a map of the body's internal state—sensations of heat, pain, and pleasure.

Together, these descriptions paint a picture of sensation as "landing" in the back half of the brain, leaving the front to figure out what to do with it. To recognize objects, for example, the visual cortex must be able to discern where edges start and end, put them in perspective in terms of far versus near, determine whether the object is in motion, and so on. Only when this type of decoding has been done can the front of your brain assign names to the objects detected and consider their relevance to you.

Neuroimaging studies have shown that attending to a particular sense modality activates a particular set of brain regions. These consistently located neighborhoods act as a collection of processing plants, quietly going about their business, each receiving shipments of raw data from its respective sense organ. The visual cortex, for example, receives the pattern of light and darkness falling on the retina. The auditory cortex receives pitch and intensity from both eardrums; the somatosensory cortex receives location information from vibration on the skin's surface.

Each sense requires its own dedicated neighborhood because of the incredibly complex task of taking noisy sensory stimuli, breaking them down into useful parts, and repackaging the cleaned-up components into usable packets of information. Each processing plant has a particular job to do, after which it sends its product to another facility farther down the road. When enough pieces are assembled, we have the information we need for what's called perception.

If you cut any of the neural wires or damage these primary sensory regions, you lose the ability to have the associated sensations. We took advantage of the fact that these regions are so well delineated in our first study together, one of the world's first neuroimaging studies of mindfulness training. Unlike activity in the front of the brain, which could represent a variety of cognitive activities, the fixed nature of sensory processing means we knew whether our participants were open to sensation. Activation of sensory regions is unequivocal; you know it when you see it.

The implication of this universality is that we all have the potential to leave abstract reasoning, judgment, and rumination behind while we channel limited metabolic resources into these well-circumscribed sensory regions. In the movie *Wedding Crashers,* Owen Wilson's character John offers the cheesy pickup line, "You know how they say we only use 10 percent of our brains? I think we only use 10 percent of our *hearts*..." While his anatomical knowledge is certainly incorrect, John's sentiment is not entirely off base in describing the extent to which we prioritize cognition and neglect sensory experience.

ACTION IN THE FRONT

Just as neuroscience has established that sensation begins in the back of the brain, it's equally certain that our responses to sensation are generated in the front. The upside-down map of your body, the sensory hairband, lies against a chasm that splits the front and back halves of the brain, appropriately named the central sulcus ("sulcus" means valley). On the other side of that sensory map is a second upside-down body map, but this one, known as the motor map, sends signals *out* to your muscles and organs. No matter how complicated your reasoning, when you decide to move any part of your body, the signal that runs down your spine from your brain originates here. So, the clearest distinction in popular neuroscience should not be between left (logic) and right (emotion), but between front and back, with *input* from our bodies and the world beyond it landing mostly at the back of the brain, and *output* to the body leaving from the front of the brain.

The motor strip, however, is just the beginning of the front portion. Before we're ready to send commands to the body, a lot of planning and processing must happen, and the brain is organized

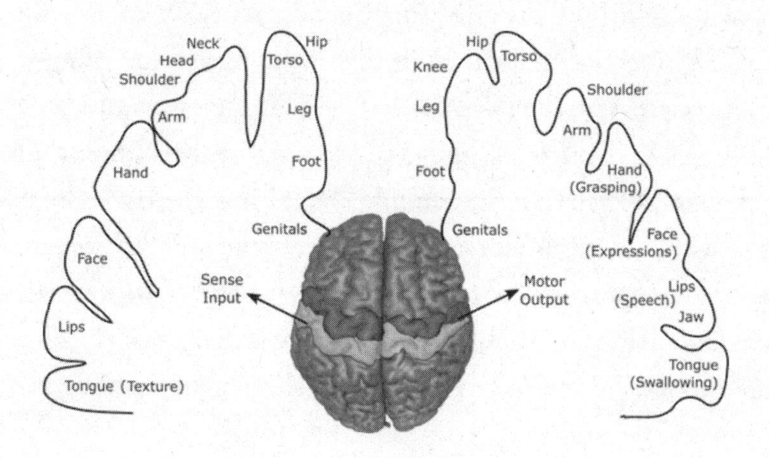

accordingly, with lots of preparatory machinery surrounding the confluence of the sensory and motor maps. Just before the motor strip, moving forward toward the face, the supplementary motor cortex plans *how* to implement behavior. And even farther forward is the prefrontal cortex, where we consider *what* actions to take given our current context. This is the seat of our distinctively human ability to consider alternatives, preferences, and consequences, the so-called executive function, which integrates life experience to inform thought and behavior. Located in the dorsal (top) part of the prefrontal cortex, this appropriately named dorsal attention network butts up against a major hub of the DMN, where habit tries to impose itself on these executive decisions.

This front/back distinction is central to the role of sensation in this account of the brain and behavior. Sensation acts as a chaotic counterweight to the DMN's imposition of order, challenging the habits that leave you stuck.

SADNESS PUSHES TO THE FRONT OF THE LINE

Segal's work on depression vulnerability made it clear that people vary widely in how brutally they judge themselves following an upset. Sheline clarified this point, showing that when people are actively depressed, the DMN often overruns its front-of-the-brain boundaries and intrudes into its neighbors' yards, impinging on other mental functions (including "thinking clearly"). But why does this become a lifelong problem for some and not for others? Are some people just born with a "bad neighbor" DMN that overreacts and stirs up trouble? Or does life experience make some people more vulnerable to the encroachment of the DMN into other brain regions?

To answer those questions, we needed to catch the DMN overstepping its bounds following an upset. So, we developed a sadness-inducing task for use in an fMRI scanner. We asked participants to watch sad film clips while their brains were being scanned. These were real tearjerkers: a cancer-stricken mother saying goodbye to her children (*Terms of Endearment*); a son reunited with his boxer dad, only to lose him again when he succumbs to fight-related injuries (*The Champ*). To control for the effects of watching film clips per se, we also measured what happened to participants' brains when they watched emotionally bland fare such as Home and Garden Television (HGTV).

As anticipated, the DMN became much more active in the brains of our sad-movie watchers. Elevated sadness ratings indicated that the film clips did their job, and the DMN did its part to contextualize and personalize the experience. Participants often left the scanner, moist-eyed, saying things like "This was just like when I had to say goodbye to my own mom."

But something curious and unexpected showed up in our analysis. We had hypothesized that more DMN activity would promote greater activation of sensory parts of the brain, a sign that people were being flooded with negative feelings. But again, confounding our expectations, we saw the opposite. In most people, sensory brain regions were *deactivated* in response to sad feelings. They simply shut down.

This threw a slight wrench into our hypothesis that it was the DMN driving depression vulnerability. There was considerable variation among participants, from those doing well to those who were moderately depressed, so we looked at whether the individual differences in the brain's response to sadness were correlated with depression scores. Here again, what we found was

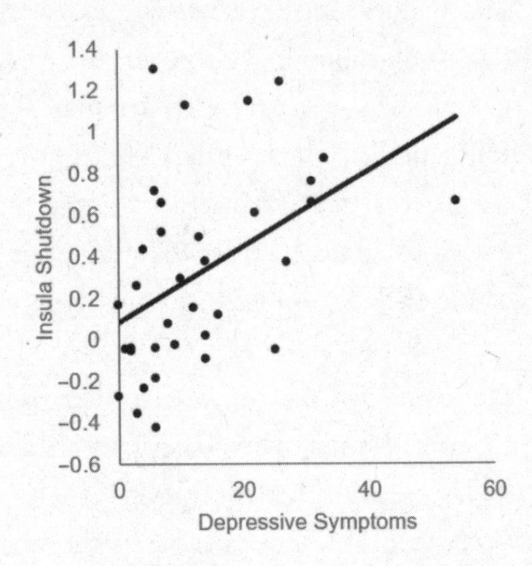

not what we expected. The contextualization process was universal and unrelated to depression, meaning that the DMN was consistently activated by sadness regardless of how well someone was doing. What ended up predicting symptom intensity was *not* a variation in DMN activity but the *extent to which sensory processing shut down.* Sadness inhibited a brain region called the insula, the "island" between the deeper levels of the brain and the higher-level cerebral cortex, and the more it shut down, the worse people felt.[2]

These results were surprising because the insula is one of the sense factories, receiving the building blocks of visceral feelings from the body—butterflies in your stomach, a pang of pain, the warm glow of a loved one's caress. The insula shutting down reflects feeling *less* rather than more emotion, because sensations, especially feelings from inside the body, are the building blocks of emotion. Our emotions are combinations of raw sensations from our bodies and our interpretation of them; to deal with their negative emotions, our study participants were shutting out the sensory building blocks that give rise to their feelings of sadness, worry, and distress. By blocking out visceral sensations, they inhibited their ability to feel badly, likely searching for a sense of relief.

Yet despite how plausible it seems that blocking out sensation would reduce sadness, the participants who blocked out sensation did *not* report feeling less sad in the moment—it's just that their sadness had become a type of expectation, concept, or thing, rather than a living, embodied experience. And this "hollow" sort of sadness was associated with higher levels of depressive symptoms—trouble with mood, concentration, sleep, and appetite. So, the sense inhibitors in our study were successful in performing a sort of alchemy in response to sadness, carving away the visceral sensations that can make up the emotion in the moment, but in doing so they had condemned themselves to a conceptual, temporally extended, and unintentionally more persistent sad mood.

These findings challenged our notions of depression as a condition of overwhelming emotional experience. When you picture intense sadness, you might envision a distressed five-year-old, tears and snot streaming down their face, totally consumed by the intensity of their feelings. But a depressed person isn't experiencing sadness in the same way—sadness in depression is often characterized by the *absence* of embodied feelings, replaced by numbness, isolation, and avoidance. Our most depressed participants had somehow learned to regulate their feelings away, an understandable reaction to being continually exposed to stress, loss, and harsh internal judgments. Yet because physical sensations are the building blocks of emotion, the more a person shut out the feeling of what was happening in their bodies, the less they could experience *any* emotion.

Here was the key to why human beings experience Selye's resistance mode. Our earlier study showed that meditation training expanded the self by including momentary body sensation. Here, exposure to negative emotions did the opposite, inhibiting momentary sensation and thereby contracting the sense of self to rely primarily upon conceptual information. But concepts are a double-edged sword because they transcend the moment and persist across time. When those concepts are negative and we inhibit sensation, there is no new information coming in to challenge these concepts, and so we are stuck with them. Judgments of being a failure or inadequate reign supreme when no new information is allowed to pervade our minds.

We knew from earlier work that people who were taught to practice mindfulness and body awareness had a far greater ability to experience the self through the senses. To see if they would respond to sadness differently, we had again used a "waitlist control" design. Half our participants had been randomly assigned to complete mindfulness-based stress reduction (MBSR) before entering the scanner. We still expected this group to show reduced DMN activity, as we had in the first study, because this kind of training helped people stop automatically following sadness down a rabbit hole of self-loathing. But once again, the action was not where we'd expected to find it. Even participants trained in body awareness showed sadness-related activation of the DMN. What made them distinct was that, in response to the sad films, the mindfulness group *didn't inhibit sensory processing* the way the untrained participants did. In the final analysis, broader symptoms of depression were predicted not by the degree of DMN activation but by the degree of sensory deactivation.

Our untrained participants spent much of their time in the scanner personalizing the experience of sadness to the exclusion of any other input. It was as if the sad movie clip had tested the limits of what they could bear, which prompted them to put up a wall against feelings. The trouble is, this numbing of sensation, rather than being protective, made their depression worse. Our trained participants, on the other hand, had learned to treat the *feeling* of sadness not as a problem to be isolated and stamped out but as an invitation to widen their perception.

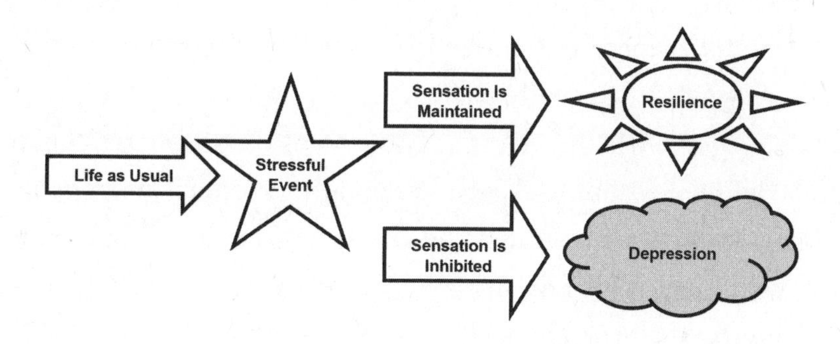

STRESS POKES HOLES IN YOUR INTENTION TO SENSE

If shutting down sensation makes us feel worse, why do these tendencies persist without our consent? The answer begins to emerge when we examine the most common strategies for managing negative emotions.

Let's say your friend snaps at you. The worst thing you can do is simply try to act normal, *suppressing* any expression of the emotion. You end up feeling not only the pain of rejection but also the pain of being silenced.

Slightly up the pecking order is distraction, which doesn't help you change the meaning of something upsetting but at least gets your mind focused elsewhere. You console yourself that maybe next time will be different; it's best not to dwell on the bad stuff.

At the top rung is reappraisal, when you figure out an acceptable reason for why things happen, a way of looking at events that reduces or removes the emotional sting. You realize your friend is having trouble at home and their reaction isn't even about you. This sets you up to reengage without having your feelings hurt.

Most people agree that reappraisal is better than distraction and suppression. The problem is that, when we're really triggered and seething, our ability to calmly reflect and make empowering reappraisals evaporates.[3] Instead, we opt for distraction to calm down, because when really challenged, we can't bear to stay mentally engaged with the stressor. Months of classroom lessons on the benefits of reappraisal are not enough to prompt most people to bring it into a truly heated encounter. Distraction offers immediate relief, but it also diminishes your chances of understanding or changing the problem itself.

We persist in this negativity because stress shuts down sensation, which makes it very difficult to see anything other than what we already know. And when what we "know" is upset and feelings of inadequacy and the struggle to cope, the DMN remains in charge, and competing perspectives that might stop the self-torture can't break through. We stay in the familiar rut, moving from stress to distress to depression, anxiety, and breakdown.

What we need is a mechanism that allows new meanings to emerge in a safe way, a way that preserves the ability to work with and resolve the challenging experiences. To learn to keep the sensory pathways open for more effective engagement, it helps to really understand what it feels like when stress shuts down those avenues.

Exercise: Noticing the Shutdown

Pretend you've just arrived wherever you are now for the first time, and, like a real estate agent, you want to assess the room's features. (If you're outside, you could be a gardener, sizing up the flora that surrounds you—or the lack thereof.)

What can you notice? What can you sense? What do you see? Does the space seem colorful? Can you hear anything? What's your state of mind as you look around? If you are sitting, can you feel your feet touching the floor? Is it hard to be curious? Try to notice whether colors, sounds, shapes, and textures feel distinct and "real" to you, versus cloudy and distant.

Now let's pause from all the sightseeing and inject a tiny element of drama. Think of a recent stressor, a situation that you found challenging and perhaps upsetting. Not an intense trauma or something really overwhelming—just one of those everyday annoyances or conflicts that tend to wear you down. Perhaps it's the state of traffic where you live, or the price of groceries, or an unkind word from a pretentious acquaintance. Sources of annoyance are not hard to find, so just for a moment, pick one and really dive into thinking about it. See if you can place this problem as the foremost thing in your mind. Plunge yourself into rumination, wondering how you can deal with the problem, wondering why you even must deal with it, and letting yourself experience again what it's like to be in this problem-facing mode.

Feeling terrible? Great! Now look around the room again and see what you notice. What's it like trying to forage for sensory information with this problem weighing on your mind? Do your senses seem more or less relevant? Does it seem easier to dismiss the importance of sensation, as if there were more pressing concerns to think about? When you canvass your senses, has anything changed? Do colors seem more "grayed out," the shapes in the room overly familiar or unremarkable, the act of feeling the texture of your seat just a pointless exercise? It's not that you can't see, hear, or touch what's around you. It's that you begin to ask, "Why bother?" This is stress-related inhibition in action. Just a little stress can rob you of the ability to care about taking in new information.

So, what is the alternative? We have too much practice exploiting our knowledge, even when that knowledge is upsetting and counterproductive. We need more training in exploration.

4

ATTENTION AS EXPLORATION

You have brains in your head. You have feet in your shoes.
You can steer yourself any direction you choose.
DR. SEUSS, *OH, THE PLACES YOU'LL GO!*

LEARNING TO BE HERE NOW

After 9/11, the Dalai Lama sent a team of Tibetan monks to painstakingly construct a sand mandala at the former site of the World Trade Center, only to destroy it as soon as the pattern reached perfection. Your first reaction might be to gasp and wonder, "What was the point? Why add a model of destruction to destruction on a massive scale?" All that effort, such painstaking detail, an intricately colored design and . . . poof! But the purpose of the mandala was *to be* purposeless. If anything, it was meant to demonstrate the virtue of every now and again letting go, giving up the illusion of control, and being purposely nonproductive.

If this seems incredibly wasteful, it may be because, from an early age, our brains are conditioned to learn new things, then profit based on exploiting that knowledge. Report cards are like

that. Performance reviews. We invest in products, not just processes. Where's the product for these monks in their sandbox?

It's certainly not the mandala itself, which lasts only an instant. To *exploit* the finished artwork for some practical purpose, or even to preserve it for posterity, would be to subvert its training purpose, which is to enable sensory engagement in the here and now. The product is *what happens to the monks* who engage in the process, along with those who watch, helping them become more sensory beings, comfortable living in the moment. Implicit here is the idea that the mind can be trained to enter a state of sensation.

In 1949, Donald O. Hebb, a founding father of modern neuroscience, described how the brain's learning and forgetting occur in each moment. Brain regions activated at the same time, he argued, forge stronger connections, making them more likely to activate together in the future. He conveyed the essence of this process, known as *Hebbian learning*, in the aphorism "Neurons that fire together wire together." Conversely, if coactivation is prevented, connections between regions decay, a process known as *extinction*.

Many decades later, we now know that brain rewiring occurs rapidly whenever there is new learning. In fact, evidence of brain growth is visible after only a week of developing a new skill, whether it's suturing a wound or juggling tennis balls! If a person can learn to juggle in a week, thereby permanently altering their brain structure, what changes might be possible when one learns to broadly engage in sensation?

Back in the lab, these questions preoccupied us as we wondered how to undo the inhibiting effects of stress on sensation. When the fat is in the fire, the DMN tries to lock us in the House of Habit, prompting us to try harder at doing what we've always done. Our hope—and our hypothesis—was that if people could be trained to pay attention to sensation, and that if new learning strengthened neural pathways, it might weaken the DMN's protective tendency, letting sensation break through with new information and opportunities.

ATTENTION IS THE THIN EDGE OF THE WEDGE

The mindfulness training some of our participants underwent helped to reduce their stress by increasing their access to sensation. This finding prompted us to take a closer look at the most basic aspects of that training—attending to the breath. We knew from past research that, with a bit of effort, people could engage their external senses and thus activate the sense factories in the back of the brain. We wondered, could paying attention to the breath similarly "flex" different neural muscles, shifting metabolism to the sensory regions?

We fired up the fMRI and asked participants to flex their attentional muscles within the sensory realm.[1] We asked people to engage in two particular types of attention—interoceptive attention (IA), tuning in to the breath or the feeling of the heart beating, and exteroceptive attention (EA), tuning in to the world around you, as when you notice the beat of your favorite song. In daily life, interoceptive attention turns inward to the butterflies in your stomach before the big presentation; exteroceptive attention helps you notice a stop sign, even when you're driving while distracted by the stress of being late for the meeting.

To test whether attention shapes brain activity, we had participants alternate between attending to the breath (IA) and viewing words on a screen (EA). The existing research literature told us where to look for the effects of attention—vision should be first represented at the very back of the brain, in the primary visual cortex, whereas the feeling of the breath should activate the bottom of the somatosensory hairband, just in front of the ears, and deeper inside to the insula, a representational landing spot for our deep-seated internal body sensations.

Our findings showed that where participants focused their attention predicted brain activity, even though some sensory processing occurred automatically regardless of where attention was directed. For example, changes in participants' breathing rates were always registered in the insula, but the insula's tracking of breathing rates became more accurate when people were paying attention to the breath. Focusing on the breath caused the correlation between faster breathing and greater insula activity to *double* in strength. During IA, the insula also showed greater connectivity to the brain's sensorimotor switchboard—the thalamus. Conversely, attending to the screen (EA) was associated with greater activation in the visual cortex, and the visual cortex showed greater connectivity to the thalamus.

In other words, you don't need an instruction manual, surgery, or months meditating on a mountaintop to rewire your brain and engage sensation to overcome habit. All you need to do is pay attention.

ATTENTION LEAVES A TRAIL

Our study showed that attention could move neural activation to different locations in the brain, but could these changes be made robust enough to hold up under stress, when the DMN is on high alert and demanding that you follow the tried and true? After all, there's no team of scientists standing behind you to remind you to attend to your senses when the popcorn hits the fan.

To explore this question, we compared the responses of two groups of participants to the same tasks used in our previous attention study.[2] Again, we employed a "waitlist control" design that contrasted recent meditation trainees with those waiting to receive the training.

As expected, attention provided fuel for sensation, moving the overall pattern of activity from the front to the back of the brain. But for untrained people, this spark of sensory activation died out before it could communicate with the front of the brain. It was like watching a stream of traffic arriving at a retractable bridge, only to pile up waiting for a large vessel to pass below. To the observer, the stream of taillights stops before the vehicles can cross the water, even though a pathway exists. Yet people with training carried sensation forward all the way to the gates of the prefrontal cortex, the seat of awareness and decision-making. Practicing attending to body sensation meant learning to keep the bridge extended more of the time.

Further, repeated training strengthened the insula pathway even when attention was absent. Attention temporarily extended the bridge to the prefrontal cortex, but repeated training was like learning to leave the bridge extended, letting sensory information reach awareness even by default. Evidently, paying attention not only shapes how our brains work in the moment but also influences the lasting structure of the brain itself. When we stop attending, the sensation fades, swept away by the metabolic current. But because *the underlying structure has changed*, we're left able to access sensation with less effort.

Given the powerful effect of sensory attention on our well-being, why don't we provide this kind of training for kids in schools? We have physical education to promote healthy body development, but we don't really have a corresponding training program to promote stress reduction and mental health. Books and computers are great, likewise basketballs and hockey sticks, but it is also important to encourage kids to notice the slant of light on the windowpane, perhaps the emerging rainbow of refracted light as the sun's angle shifts over the day. Perhaps they could learn to see their own thoughts and feelings as shifting landscapes rather than fixed properties as well.

The popularity of yoga and mindfulness shows there's a hunger for reuniting with the world of our senses, but many people remain skeptical, seeing something quasi-religious in these meditative practices. Developments in neuroscience, however, should provide reassurance that doing sense training isn't any more mystical than doing a few stretches in the morning. We understand that muscle tightness comes from a lack of fiber mobility, and so stretching makes sense. Similarly, incorporating greater sensory awareness into our daily routines is key to keeping the brain's sense "muscles" limber, letting them flex and stretch and reach the front of the brain. "Stopping to smell the roses," as it were, ceases to be a truism easy to dismiss and becomes a vital habit we can learn to access even when the deadlines pile up and the stress mounts. When stress cramps sensation, making time to forage within our senses helps to relieve the tension.

PAYING ATTENTION COSTS US

To understand why it is so important to build these sensory muscles, consider the following thought experiment. Imagine that your entire life, from cradle to grave, is contained within a long tunnel of your experience. Each cross section of the tunnel represents a single moment in time; it contains all the different elements of your experience available to you:

- Your external senses (vision, hearing, taste, touch, and smell)
- Your internal senses (feeling hot or cold, hungry, sleepy, pain, pleasure, and perhaps the feeling you get after too many late-night snacks)
- Your thoughts—the internal narrative playing out in your mind in each moment
- Your memories—the past experiences and knowledge that you are calling to mind

- Your expectations—your imagined future, including your hopes, wishes, and fears

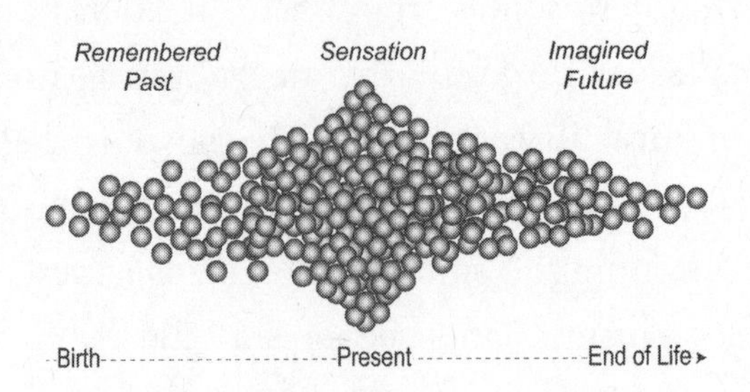

Now, your attention is a limited-capacity system—there's only so much attention to go around. When something captures your attention, there may really be no ability to notice other elements of experience. This does not mean that you can't be aware of multiple things in a given moment; in theory, you could spend your time scanning across the different elements, moving from senses to thoughts to memories and back in a constant scanning of your present experience.

But in practice, that is not what happens. We notice some aspect of our experience, a thought or sensation, that is salient because it connects to previous experiences and the anticipated future. So, while attention *could* freely roam across a broad range of experience, what *happens* is we fixate on only one or two elements of experience because most of our attention is spent integrating these elements into a narrative. We can't help but create a story that makes sense of some salient thought, feeling, or sensation. But the consequence of such storytelling is that there isn't any attention left to notice the many other elements present in any given moment.

There isn't a problem with this tendency most of the time, when our narratives create a sense of meaning and purpose and help us to focus on the most important things in our lives—we benefit from the efficiency of attending to a friend's face when they are upset, a traffic light at an intersection, or a work project as a deadline looms. Yet when our attention habits don't serve us, we are blind to alternative options because attention is stretched so thinly across the present as it extends into the past and future. We see each moment as though we were seeing the whole picture, but we are deeply conditioned to be aware of only a few facets of that picture.

This is precisely why rebuilding sensory pathways is so transformative. Because attention is a limited resource, a powerful way to break down a dominant narrative or worldview is to starve it of those attentional resources. A simple way to do this is to attend to elements of experience that are not already bound up in this worldview—every moment of attending to a sensation that is not obviously relevant is a moment of taking resources away from the story we tell and reengaging with the unknown. Over time, the brain's property of extinction will reduce how easily a story comes to mind, opening the door for new narratives to emerge. This creates a sandbox of sorts in which we can choose between the stories that run us, so long as we can put the work into attending to our experience in new and exploratory ways.

In thinking about the "tunnel of experience," we can see how a walk outside can be just another walk, as we attend to the same old things and repeat the same familiar stories. But *the next walk you take can change your life—if* you understand how paying attention to the unfamiliar elements in your experience will weaken the grip of your dominant narratives and open you up to discovering something new.

Exercise: Foraging for Change in Sensation

As we established earlier, one of the brain's most basic functions is to take random and noisy input from the senses and repackage it in ways that create order and meaning for us—which raises the question of how much of the input is dictated by someone else, whether a boss, a social group, a media company, or an advertiser. The good news is that you have a say in what you attend to, regardless of what others have in mind. Even on a busy day, peppered with distractions, your senses are still available for exploration. Our studies show that you can reclaim where you want to focus, rather than leaving your mind at the mercy of the attention economy.

Just by spending more time engaging in sensation, you can strengthen the attention pathways for sensory awareness. By observing how familiar and seemingly static sensations might really be made up of lots of changing parts, you are getting better at moving past only seeing what you expect to see or relating to your body in the same way. Our research shows that when you are practicing attending to your senses, you are probably activating parts of your brain that tend to be neglected. If you keep activating these regions, you will start to form new neural pathways, literally changing how your brain is structured. The more time you spend in sensation, the more neuroplasticity will make sensation available to you, like adding a new friend to your contacts list (or "speed dial" for those of us who still know what a dial is).

But what happens when you set aside the familiar names we assign to sensations and treat them instead as dynamic forms of energy being received by our bodies? Because energies such as light or sound have their own ebb, flow, and time signature, you can be more fully immersed in sensation by focusing on these particular features.

1. **Change and Movement in Seeing.** Find a wide-open space to look at. (A window will do if you're indoors.) Take a moment to have a look around you, up and down and left and right. Now select one object in your field of view, near or far, and start to focus on the visual elements that stand out to you. For example, if you see a tree, notice any movement, parts that are darker and lighter, smooth and rough patches, contact with the ground and asphalt, symmetry and asymmetry, shadow and sunlight, and whether these elements stay the same or change in any moment. You might also notice the urge to name what you're looking at (bark, branches, trunk, leaves), but try to resist. Try to experience these objects in a purely visual way.
2. **Change and Movement in Touching.** One of our most common but commonly overlooked sensory experiences comes from contact with our clothing. For a moment, focus on the fabric touching your skin, the feel of each different texture. Do you notice places where there's tightness, bagginess, roughness, or softness? Do these sensations stay the same or do they change when you walk or sit? Once again, try to suspend the cognitive labeling activity and focus on the experience purely through your tactile sense.
3. **Change and Movement in Hearing.** Whether indoors or outside, stop for a minute and just listen to the sounds around you. See if you can notice how sound is layered, with the loudest and most familiar sounds coming through first: traffic, voices, running water, doors opening and closing, phones ringing. Then listen for sounds that are below that layer, perhaps your knapsack bumping against your body, your breathing, fingers touching a cup, or cutlery

on a table. Notice the qualities of each sound, whether it's near or far, intense or subtle, continuous or intermittent, surprising or expected. Notice also the space between one sound leaving and another arriving, as well as those moments of silence, when there is no sound at all. Once again, try to stay with the qualities of the sounds themselves instead of labeling what you're hearing.

4. **Change and Movement in Smelling.** Find five things around you, some familiar and some unfamiliar, and really smell them. These things could be coffee grounds, a bath towel, cooking spices like cinnamon, ginger, or curry powder, a bar of soap, boiling water, potting soil, your underarm, wet leaves. This time as you inhale, instead of putting a label on the scent, see whether you can describe its qualities, and whether it changes or stays the same over time. Is the scent intense or subtle, does it arrive quickly or slowly, does it register surprise, is it continuous or intermittent, does it fade into another scent or stay the same throughout? Do all the scents trigger memories or recollections for you, or only some?
5. **Change and Movement in Tasting.** The next time you're going to have a snack (maybe eating one or two things instead of a meal), see if you can make time for this exercise. Instead of naming any of the primary tastes, simply register a taste or tastes as present and describe its/their qualities and whether they change or stay the same over time. Is the taste intense or subtle, does it arrive quickly or slowly, does it register surprise, is it continuous or intermittent, does it fade into another taste or stay the same throughout?

After any of these exercises, it might be a good idea to check in with how you're feeling. Having extended some of your sensory bridges, did you have any experiences that surprised you or were unexpected? How did it feel to let new information in? Was it exciting or perhaps a bit anxiety provoking as you entertained the unpredictable nature of your senses? Regardless of whether you enjoyed the experience, did anything open up for you in terms of how you responded to your sensations? Did you change your mind about your preference for a smell or taste? Did you experience your body or the world around you differently?

At this point it is totally okay and maybe even a good sign if you aren't sure what to take away from these experiences—by sensing the world on a moment-by-moment basis, you are getting a taste of what it's like to value change over certainty. It can be deliciously unsettling to value the uncertainty of sensation in a world that valorizes certainty and action.

5

AN ALTERNATIVE TO ACTION?

Today, like every other day, we wake up empty and frightened. Don't open the door to the study and begin reading. Take down a musical instrument.

—RUMI

ON JANUARY 15, 2009, CAPTAIN CHESLEY "SULLY" SULLENBERGER saved 155 lives by safely landing US Airways flight 1549 in the frigid waters of the Hudson River. Having lost both engines shortly after takeoff, he was in uncharted territory. His first thoughts, as he recalled, were "This can't be happening . . . This doesn't happen to me."

This was a veteran pilot with more than forty years' experience. But he quickly realized that life had just violated every expectation and routine, and that his long-practiced and seemingly essential habits were not going to get this plane safely back to earth.

Despite extensive training for emergency landings, Sully had never landed a commercial passenger jet on water. Then again, who has? So, he had no choice but to improvise, find entirely new tactics, and, of course, get it right the first time. He also realized that under such life-and-death pressure, there was no room for anything but total engagement in the moment, which meant total attention to what his senses were telling him. He described his mental focus this way: "I never had any extraneous thoughts in those few seconds. . . . I didn't allow myself to and I didn't have any inclination to. I never thought about my family. I never thought about anything other than controlling the flight path and solving each problem in turn until, finally, we had solved them all."

As for his emotions, he made no pretense of being the stoic hero with ice water in his veins. In fact, as he recalled, his complete engagement brought him into closer contact with his very real terror: "[Training] develops in you an ability to summon up from somewhere this ability to create a sense of calm, a professional calm . . . but we weren't calm. We couldn't be calm. The stress was too intense, but we had to focus . . . to be able to do the job despite how stressful it was."

Instead of striving to remain detached or coldly rational, Sully placed himself entirely within the chaos of the moment, at the nexus of sensing and responding as events unfolded, setting his old routines aside and letting the stress flow through him. But wouldn't his rampaging emotions impair his ability to reason? Apparently not, as evidenced by the 155 lives he managed to save.

The key to understanding Sully's success lies in his statement "The stress was too intense, but we had to focus." What he's saying, essentially, is that there simply wasn't time to include "feeling stressed" on the list of problems to be solved. His job was not to make himself feel better but to land the plane. "Dealing with" his emotions would have led him to divert critical mental resources into self-soothing and emotion regulation. When the engines died, worrying about conforming to cultural expectations of a stoic and unflappable hero-pilot was a frill he couldn't afford.

Earlier, we discussed how many of us inhibit negative emotions when we're feeling sad. We described how we saw this pattern in the fMRI scanner, finding that the more people suppressed their feelings, the more depressed they seemed to be. Those with training in engaging sensation, however, had learned to allow the emotional pull of sad movie scenes to be processed with the same sensory throughput they afforded neutral, unchallenging HGTV clips. In other words, blocking out sensation is something we're conditioned to do, but like Sully, we can get beyond that conditioning. When we give in to our default and suppress what we feel, we often distract ourselves from what's most important. We trade away the opportunity to take in new information for the sake of meeting expectations.

Sully's first step toward saving a plummeting airplane was to *accept* the concurrent "problem" of his emotional state, that is, his feelings of overwhelming stress, his failure to be that unflappable pilot. But how can accepting negative emotions reduce their impact?

To get to the answer, it helps to understand why stress captures us so fully in the first place.

THE PREDICTION MACHINE THAT IS YOU

Over the years, various metaphors, often reflecting the dominant technology of the time, have been used to describe how our brains work. The ancient Greeks thought of our mental apparatus as a kind of theater; Freud relied on hydraulics to describe a mind that functions via the buildup and release of psychic pressures;

cybernetics portrayed the brain as a computer; self-help gurus treat the brain as a muscle, extolling "use it and make it stronger" strategies.

Nowadays, neuroscientists use the metaphor of the brain as a *neural network:* a complex tapestry of connections that match sensory inputs to motor outputs. From this perspective, your brain is *always* trying to predict what will happen next, keeping you prepared for whatever curveballs life throws at you.

Emotion researcher Lisa Feldman Barrett describes the brain's predicament as follows: it sits in a dark box (the skull), receiving only sensory clues to what is happening in the body and the world beyond. The brain continuously receives input from its sense factories: flashes of light, sudden sounds, a bitter taste, a pungent smell, or feelings of pressure or heat. The true origins of these sense packages are unknown, but each sense package suggests that *something has changed*, and the brain must determine if these changes are survival relevant enough to warrant a response.

Fortunately, our brains have evolved to rapidly catalogue each sense package, comparing it to past experiences to infer causes and implications. This categorization allows the brain to fulfill its chief mission: to curate reactions that keep the body alive. As Feldman Barrett reminds us, the brain did not evolve for logic, joy, or accuracy but to regulate the body that houses it.[1] The most efficient way to accomplish this, especially when the action heats up, is by anticipating the body's needs and satisfying them before things get out of hand.

If you've ever gone for a walk in a park at dusk and seen a large loping animal with a coarse coat of fur, your brain immediately generates a cascade of guesses, rather than waiting for complete information about what sort of beast is coming toward you. If there have been coyote sightings in your neighborhood, you might quickly retreat and keep looking over your shoulder. If a woman walks up and attaches a leash to the animal, then the coyote prediction—and subsequent fear—fades away.

Our brains are constantly creating momentary snapshots of reality to get ahead of events and prepare our bodies to act as needed. These snap judgments are often wrong, but the alternative would be worse: taking longer to size up the situation might mean that your response is too late.

Thinking about life from the brain's perspective provides us with a different way of understanding human motivation. Imagine what it's like to be a brain trying to navigate a rush of inputs to keep its body alive. As discussed earlier, the brain must represent and respond to changes both within the body and in the environment beyond. But how does it know what it's dealing with?

From a neuroscience perspective, the brain is constantly making educated guesses and then dealing with the consequences. Modern science has a name for this framework: *the predictive coding model of the brain.*

This model is important because it aims to explain how we understand all situations, not just those involving potential predators in shadowy woodlands. Looking around my room, I see walls, a desk, chairs—but only after lots of visual impulses are packaged and processed by my brain. Light hits my eyes and is converted into electrochemical signals. My brain then sends these signals to a multitude of smaller brain factories that find edges, discern colors, or detect motion. These details are then put together in two different ways.

Along the bottom of the brain, the objects are compared to what I already know about the world, assigning names, meanings, and related memories. A separate stream of information moves toward the top of my brain to place the objects in space relative to my body, allowing me to approach, avoid, or manipulate them.[2] All this packaging is happening all the time, even when all I'm looking at is a chair or desk. When I see these mundane objects, my brain tries to match the relatively fixed sensory process of finding edges, shapes, and textures within the knowledge base of what combinations of edges, shapes, and textures are called—they have names such as "table" or "chair"—and how each can be used. But this simple act of seeing and identifying really isn't so simple.

In *Thinking, Fast and Slow,* Kahneman suggests that we should avoid making errors in judgment by relying on deliberation (System 2), rather than succumbing to the "laziness" of DMN-informed intuition (System 1). But if we should be cautious about snap judgments or knee-jerk reactions, what about snap perceptions? We may not mistake tables for chairs, but what happens when we've automated more ambiguous cues, like distinguishing a smirk from a smile, a compliment from a dig, a yawn from a snarl?

We take it for granted when the system for automated perception is intact, but whenever the supporting brain regions are damaged, we can see the complexity behind visual and other sensory categorization. Oliver Sacks's *The Man Who Mistook His Wife for a Hat* gives a humane portrayal of people living with various forms of agnosia—the condition of being unable to decipher sensory information. His case studies show how central the integration of sensory information is for everyday functioning. A person with agnosia can't identify household objects even though touch is normal, recognize familiar faces even though vision is normal, or distinguish speech from nonspeech sounds even though hearing is normal. The factories are still producing sense packages, but without some meaning attached, sensations are untethered to feeling, memory, or potential responses. Clinical cases of agnosia (along with dogs mistaken for coyotes) reinforce the idea that there's nothing special about a table, chair, tree, or cloud that imbues it with its name and function. We need the brain to correctly identify and label things before we can have any inkling of how to interact with them.

Yet naming a thing isn't enough. Once the brain knows what it's dealing with, a second question arises: Does this matter to me? Categorizing a sensation helps describe the physical characteristics of objects but says little about its significance. *You* are the one bringing meaning to each of these sensory experiences. From the brain's perspective, each moment of recognition must be paired with a more primitive inference derived from eons of evolutionary history: is this relevant to my survival?

Our brains are constantly trying to *make sense* of sensation, to transform the chaos into something predictable or expected, and much of our knowledge is learned over time. Imagine a parent pointing to a whistling kettle and saying to a child, "Kettle. Hot. Makes tea. Don't touch!" The child's brain not only has to categorize that this object is a kettle; it also has to learn to attach to this category instructions for cautious interaction because within the chaos of sensation lurk real opportunities and real dangers. This tagging of categories and then self-relevance usually happens automatically, enabled by something going on behind the scenes that gets us ready to recognize the world.

Understanding that hidden "something" lies at the cutting edge of modern neuroscience. According to *predictive coding* theory, our brains don't have time to size up each new situation completely from scratch, so they continuously predict the next moment of our reality, then check to see if the prediction is right. This capacity, honed through millions of iterative forecasts, confirmed or disconfirmed, is why you don't have to examine a room the fortieth time you enter it. It's also why you expect certain things (table, chairs) but not others (ducks, waterfalls) to be in your kitchen. Sensation becomes a prediction game, and one that we can usually win without even noticing that we're playing it.

YOU ARE WHAT YOU PREDICT

We walk around assuming we perceive the "true" world around us. However, even our most basic perceptions may not be universal but largely idiosyncratic. Anthropologists have long known that what we care to categorize is culturally specific, with predictions made to the extent that they help us navigate our specific environment. In *Tristes Tropiques*, Claude Lévi-Strauss described the classification systems used by certain tribes in the Amazon rain forest. Instead of mental categories devoted to animal, vegetable, or mineral, they group "things that belong at the top of trees," "things that sit under trees," or "things associated with rocks." Presumably, where to find things amid the vegetation matters a lot to these tribes, while a formal scientific classification system would be of no use. Other folk taxonomies have emphasized the fruiting patterns of trees or the behavior patterns of large animals that relate directly to tribe activity and survival.

For many of us, snow is just wet white stuff that we need to wipe off our windshields before we can drive, but for the Inuit in Northern Canada, finer distinctions and vocabulary emerge. These hunters from the frozen north have about forty words for snow, distinguishing, for instance, between aniu, snow used to make water, and maujaq, a snowfall in which a person might sink. Swedish people distinguish between aprilsnö, spring snow that may indicate flooding, and kramsnö, which is good for snowballs. More specific predictions emerge in both of these northern cultures because knowledge of snow is important for survival.

Each of us has our own cultural and survival priorities, and our brains filter through countless possibilities to arrive at a meaning that makes the most sense of what our senses are taking in. The overriding objective is to reduce *prediction error*, which is the gap between expectation (our prior knowledge of the world) and sensation (the information coming in about the present moment).

The predictive coding model of the mind may have been first introduced by the philosopher Immanuel Kant in the 1800s,[3] but it's become far more sophisticated and central to cognitive science through the centuries. It suggests that *all human motivation* comes from trying to minimize prediction error at all levels of the brain, generally without our awareness.

At lower levels, reducing prediction error might look like the brain figuring out where a door should appear in your visual field. At higher levels of cognition, the stakes are raised, and minimizing prediction might concern the operation of the entire organism. This is where motivation comes into play, as higher-level predictions might include whether the door should be locked, cuing concerns around safety. As we mature, most of us get good at feeling secure in our environments, a sign that we're successful at resolving prediction errors, minimizing moments of confusion and uncertainty, and no longer worrying about monsters lurking under the bed. Sooner or later, though, we find ourselves in unfamiliar territory, and the need to deal with surprise and violated expectations rears its head once more.

WHAT YOU PREDICT IS WHAT MOVES YOU

The drive to minimize prediction errors is motivational because the central axiom of the model is that discrepancies lead to action. If you wake up and everything seems blurrier than usual, you might rub your eyes, or you might just remember to put on your glasses. If the blurriness persists, you'd very likely go see an eye doctor. When you step onto a floor you predict to be level and smooth, then feel the sharp edges of a Lego block jabbing into your foot, you immediately reduce the pressure on the foot (ideally before deep tissue damage ensues). When you enter your house and notice a strange smell, you might suspect a gas leak and seek help. When you look at the face of a friend and don't see the expected light and engagement in her eyes, you might assume that something's wrong and be motivated to help. You might even predict feeling nothing in particular as you go about performing the same job day after day, and then be surprised to discover a deep well of unhappiness or resentment, after which you feel motivated to find other work.

Again, the reason we feel surprised or upset is because this prediction architecture is trying to orient us to the changes needed for our survival. The most typical response to failures of prediction is *active inference.* We interpret the feeling of surprise (the prediction error) to mean that something is wrong in the world and that we need to take action to fix it. This is the classic Western method for bringing prediction and reality into alignment.

Many great achievements come from a violation of expectation that's met with active inference. Take Alfred Nobel, for example, the nineteenth-century inventor who went on to fund the prizes that bear his name. During his lifetime, he was known as the inventor of dynamite, a man whose discoveries enabled major advances in engineering and construction. But he was also known as one of the first and most notorious global arms dealers, a man who provided weapons to fuel the arms race of the late nineteenth century. A French newspaper gave him the rare opportunity to consider his legacy when it mistakenly reported that he'd died. In fact, it was his brother Ludvig who'd passed on. But presented with this rather extreme violation of prediction—he knew himself to be very much alive—Nobel was momentarily sensitive to how he would be remembered. To his horror, the obituary reported, "The Merchant of Death Is Dead," and condemned Nobel's discoveries, saying that they had led to wide-scale suffering. Nobel was appalled, and luckily, he still had time to act. In short order, he restructured his will to leave 94 percent of his immense personal fortune to establish the Nobel Foundation. This lofty prize has likely motivated more than a few of the world's brightest minds in subsequent years. So out of the upset of being surprisingly (at least to himself) cast as a villain, Nobel was driven to *act upon the world* to reestablish his narrative of being a good and valued contributor to society.

Active inference is how we're taught to operate most of the time as adults, and it's a microcosm for modern society. When the world violates your expectations, you try to change the world, and in many cases, this is a great strategy. A Lego piece *ought not* to be embedded in my foot, and I should take it out and examine the floor for other pieces I might step on. My friend *ought not* to look so morose, so I should do something to cheer her up. My job *ought not* to leave me feeling so unexpectedly resentful and unsatisfied, so maybe it's time for me to leave the job. When we're able to act to reduce the error between prediction and sensation, we experience *agency*, the feeling that our actions make a difference.

From a psychological perspective, believing that you're captain of your own destiny is a huge piece of the wellness puzzle. Whether it's surrounding yourself with people who support you, excelling in sports even when the odds are against you, or tolerating the effects of chemotherapy to recover from cancer, certain actions confirm the belief that we can play a role in determining our own fate.

Scott Hamilton, a US Olympic gold-medalist figure skater, calculated that he'd fallen 41,600 times over his career. "But here's the funny thing," he said. "I got up 41,600 times. That's the muscle you have to build in your psyche—the one that reminds you to just get up."[4]

It would be easy to dismiss Hamilton's advice simply as hard-ass discipline—he's an Olympic athlete after all. The lesson for us, though, is that falling and getting up taught him to get beyond a fixation on his failures as separate instances and to see them instead as an essential part of the larger journey he'd signed up for. In that way, falling reminded him that he was still on the journey, and getting up was simply the next step in that progression.

Just like Hamilton, when each of us engages in active inference, we're throwing the dice, attempting to claim some of that sweet agency, to restore a sense of order by shaping the world around us to fit our ideals. However, each time we attempt to create a sense of agency by changing the world, we also run the risk of failure, and there are times when it might be unwise to try. If after watching Scott Hamilton, for example, one of us decided to do a backflip on skates while moving at high speed, we would likely not experience a huge sense of mastery and agency. Agency is lovely, but what do we do when taking control and reshaping our world just aren't in the cards?

ACTIVE INFERENCE IS GREAT (EXCEPT WHEN IT ISN'T)

Active inference—signing up to solve the problem—is great when we touch something hot and must remove our hand from the stove, or when we start getting cranky because our blood sugar has crashed and wisely eat a sandwich. Yet we know that sometimes it fails, and that's when we can feel stuck and hopeless. We're aware there's a problem, but our best strategies to solve it don't gain traction.

Sometimes this happens when we try to use active inference to solve problems it cannot address. A classic example of this is when we wish we could change something that's already happened. "Why did my father love my brother more than me?" does not describe a situation we can alter through any overt action. By the same token, we can never resolve things that have yet to happen—and maybe never will. That's why asking "How can I make sure nothing goes wrong on my vacation?" is simply a recipe for anxiety. Asking "Why can't I ace my serve like I used to?" is similarly useless, because active inference has no effect on the aging process, and resenting it is a recipe for bitterness. No amount of striving to change the world will erase the past, guarantee the future, or alter the laws of nature. But even with more mundane issues, believing that mastery is our only option leads to frustration, locks us into resistance, and, as we saw with Shanice the entrepreneur, might even lead to a breakdown.

Indeed, many mental health disorders are examples of active inference gone awry. The idea of *learned helplessness* comes from research on depression; it describes when a person has so often failed to solve their problems through active inference that they begin to see their problems as unsolvable. Animal studies from the late 1970s replicated this pattern. A rat will swim frantically when dropped in a water tank to find a safe platform. But if that rat keeps getting thrown into platformless tanks, it will learn not to bother swimming and exploring. The tragedy is that the rat will continue with this defeated attitude even when dropped into a new environment where a platform is available; it will drown before it engages in exploration again.

We may like to think we're smarter than the average bear, but we're no better than lab rats when it comes to overcoming a history of active inference failure. Learned helplessness can explain why a person with depression may devalue the resources and supports that are close at hand. There's little point to getting out of the house, going for a run, or calling a friend if you've concluded it isn't going to make a difference anyway. A person suffering from depression literally can't see the possibilities around them because their future has been foreclosed by a history of active inference failures.

THROUGH THE LOOKING GLASS: REFLECTIONS FROM A THINK TANK

On the other hand, blind faith in active inference can be a type of trap, especially when it comes to our mental health. When it doesn't work, it can feel like *Star Wars*' Obi-Wan Kenobi despairing at Luke's potential descent into the dark side of the Force—"active inference was our last hope." Yet we must channel our inner Yoda and take to heart his revelation that "no . . . there is another." Yoda reveals that the "Chosen One," Luke, has a twin sister, Leia. In the predictive coding world, active inference too has a secret sister: *perceptual inference*.

This "New Hope" (yes, another *Star Wars* reference) is essential because there are times when active inference is almost certain to fail. It's simply not possible to will yourself out of depression or obsessive-compulsive disorder. Instead of trying to make the world conform to our expectations, perceptual inference allows us to update those expectations. It's often when a person gives up notions of absolute agency that they can make progress. As renowned psychotherapist Carl Rogers once said, "The curious paradox is that when I accept myself just as I am, then I can change."[5]

In the spring of 2013, Norm was invited to attend a think tank on *interoception*, the sense of the body's internal state, at the University of Massachusetts Amherst. The conference was sponsored by an organization called Mind and Life, whose mission is to advance scientific exploration of ancient contemplative techniques. Attendees were an eclectic mix of researchers in clinical neuroscience and those studying meditation, yoga, and qigong.

By combining these diverse perspectives, the group sought to bring balance to a Western medical literature (predictably) dominated by active inference accounts.[6] From a medical perspective, our feelings of sickness and health come from the felt sense of the body, also known as interoception. From a contemplative perspective, this kind of body awareness serves as a foundation for seeing one's mental habits more clearly. Instead of relying on ideas about the body, observing sensations from moment to moment sometimes tells a very different story, one of fluctuation and novelty rather than permanence and repetition. In the Western tradition, interoception makes us better agents in fixing problems and making decisions. But in Eastern traditions, reducing our need for agency—the expectation that we can remain in control—provides not only relief for many ailments but opportunities for insight as well.

The think tank sessions began with a discussion of predictive coding as a way of understanding interoception. For Western science and medicine, the point of interoception (and sensation in general) is to promote *active inference:* when we feel a pain, we should try to relieve it. The point of recognizing the unexpected in these messages from our bodies is to empower our conscious minds to recognize, categorize, and fix imbalances in our physiology. And medicine is successful to the extent that it can stop your body from sending you these surprising and unwanted signals. Psychiatry is successful if it can stop you from having overwhelmingly bad feelings. Chronic pain treatments are successful if the doctors can titrate enough painkilling medication to dull sensation without spiraling you into addiction or delirium. Even palliative care, the arm of medicine that deals with end-of-life experience, is about actively removing sources of distress and making people comfortable as they pass on.

For those coming from contemplative traditions, however, an unexpected sensation is also an opportunity to change ourselves rather than the world. Instead of trying to get rid of a source of surprise, we might explore or *forage* for an understanding of what led us to be surprised in the first place, to work with our own expectations rather than resent their violation. It turns out that the researchers who'd initially proposed the predictive coding model of the brain had come up with their own term for updating our expectations in response to surprise. They called this behavior *perceptual inference*, and it provides an alternative to always having to fix the world to justify our expectations. Perceptual inference gives us another gear to shift our minds into, one that doesn't require us to just push through when we realize circumstances aren't what we thought.

At the start of life, all we have is perceptual inference. Perceptual inference is how a baby learns to recognize that objects exist and are separable from one another. In a visual field that's just a mishmash of color and texture, nothing is predictable because nothing is clearly identifiable. But breaking down the visual field into discrete entities—a bottle, a teddy bear, a ball—dramatically reduces the chaos. We're no longer surprised to continue to see those same demarcations as our vision shifts with each movement. We begin to learn about object permanence, that there really are different things out there in the world. And every time we update our expectations to minimize the unexpected, we're engaging in the largely unconscious process of perceptual inference.

Yet even though perceptual inference is intrinsic to a child's making sense of the world, scientists have largely ignored its role for higher-level decision-making in more mature organisms. Once we know there are tables and chairs, nourishing foods and poisonous snakes, what's the point of continuing to update our expectations—to refresh our understanding of the world around us? The main thing, so the dominant line of Western thinking goes, is to shift into active inference: My perceptions tell me how I should act upon the world to reduce my surprise. If I feel a pang in my belly, perhaps I should eat. If I feel confused, I should ask a question and try to learn more. If I feel hurt by someone, I should confront them or else avoid them. All in all, in response to the unexpected, I should bring things back in line with what was previously expected. I should act to assert control, to shape the world, including my body's internal state.

A major insight of the think tank was realizing how neglected but vital perceptual inference is for a healthy adult life. While neuroscience models correctly acknowledge that surprise can be reduced through active inference, perceptual inference can provide another high-value option for responding to the unexpected. When I feel the urge to eat, I might question if I'm truly hungry, or simply bored or anxious, and whether the feeling will abate on its own. If I'm confused, I might sit with the feeling, calm down, and see whether my understanding will improve. If someone hurts me, I might reflect on how we ended up in this situation, then explore a range of alternatives, and only then decide whether to act. In other words, there's plenty of room for us to update our labels and categories, particularly for less objective things, such as the feelings we detect inside our bodies, or the intentions and motivations of other people.

It is not only a blind spot of neuroscientists; our culture, by and large, fails to see perceptual inference as a useful approach to navigating everyday life. Instead, we see our attitudes and feelings as real objects, like tables and chairs, and once we categorize ourselves as feeling sad, or angry, or worried, we're motivated to act on the situation so that we no longer have to feel that way. We rarely consider that our expectations themselves may be what need to change.

There's nothing wrong with having adaptive strategies and a good read on our feelings. It's a good idea to eat when you're hungry, to speak out when there's injustice, and to avoid nasty people and dangerous situations. But other times, when we try to fix the world to reduce the incongruity of unexpected events, we continue to feel surprised, distressed, and violated, and our actions don't resolve the discrepancy. We yell at someone for getting in our way, and we end up feeling even worse for acting like a jerk. We cram snacks into our mouths, and we still feel the restlessness we'd interpreted as hunger. In these situations, the continued violation of our sense of how things *ought to be*, even after we follow active inference, is a sign that our expectations need to change. Calling someone else's actions wrong, or telling ourselves that we ought not to feel the way we feel, isn't the exit point from this vicious cycle. To resolve the situation, we need a better model of what's actually going on. As long as we're running an outmoded model, we'll keep finding ourselves unpleasantly surprised to be plodding along in the same rut.

The failure to value perceptual inference in modern Western societies may stem from confusion about how to use it. The bias in Western culture is to denigrate refining one's perceptions as capitulation or resignation. Yet the art of questioning, of negotiating with our representation of the world, lies at the heart of the contemplative practices from other traditions. If I carefully examine my sensations, what are the alternatives to my most automatic interpretation? The value of this was already shown in the scientific literature—people who practice this art of looking again, of foraging for new meaning in sensation, are often healthier and happier. The sweet spot, given the requirements of the workaday world, is to balance active inference with perceptual inference. But first, we need to learn how to do it.

PERCEPTUAL INFERENCE: OVERLOOKED YET EQUALLY POTENT

Perceptual inference begins with valuing sensation. As children—and this is true of anyone of lower power or status—we're expected to engage in perceptual inference when our models of the world are shown to be out of sync with the world as it is. Getting to have birthday cake for breakfast every day is not a reasonable expectation, or so we're told. To the extent that we are told *why* eating cake for breakfast can lead to downstream health problems, we may grow up to appreciate that not eating cake for breakfast turns out to be good advice and that we need to set boundaries around our desired sensations to be healthy. On the other hand, a caregiver who simply snatches cake away and says "No, because I said so" achieves the same outcome in controlling the child's behavior, but to make sense of this experience, a child might internalize the notion that strong people get what they want, and strength means getting to be the person who decides what to eat. In both cases, the child must make sense of not getting what they expected to get, but the perceptual inference powerfully determines their model around future eating experiences.

The problem is that we don't often get the ideal experience of updating our models in a way that we can accept and say yes to. Much more often, we are forced to accept the reality of situations we can't change, without letting go of our resentment at our lack of power. This is a kind of model updating, a perceptual inference, but it isn't one that addresses the violation of expectation—we still wish we could change the world to our liking, and our inference is one of our own lack of power to do so. And this is why as we become adults and accumulate social standing and autonomy, we associate perceptual inference with this early experience of helplessness that we are not eager to reinhabit.

Yet even if we can learn to accept the need to sometimes update our models of the world rather than resenting such change, perceptual inference can feel like a heavy lift, especially when the cost of updating our mental model includes violating culturally sanctioned expectations or predictions. Altering your expectations can be the right thing to do, and still the outside world may interpret it as being steamrolled, bullied, or oppressed or as simply giving in. And yet there are those rare individuals who engage in perceptual inference despite enormous pressure to do otherwise.

Ash Barty was only twenty-five, and the number one women's tennis player in the world, when she announced her retirement from professional tennis. Most athletes hang in there until their bodies give out. But Barty, the first Australian woman in decades to win both Wimbledon and the Australian Open, at the top of her game and with no major injuries, simply walked away. Given the lucrative career undoubtedly still ahead of her, this might seem like a terrible decision. But her reasoning makes perfect sense:

> "Success for me is knowing that I've given absolutely everything that I can," Barty said. "I know how much work it takes to bring the best out of yourself. I've said it to my team multiple times; I don't have that in me anymore. I don't have the physical drive and the emotional want and kind of everything it takes to challenge yourself at the very top of the level anymore. I think I just know I'm absolutely . . . I am spent. I know physically I have nothing more to give, and that for me, that is success."

Barty had always been inner directed, with a drive that in some ways transcends the external competition. Part of Barty's rise in the tennis world came from living with this intense competitive feeling inside, a sensation of "emotional want," in her words, that motivated her to undergo the almost unimaginable level of discipline and sacrifice required to stay at the top of the game. As long as she could tap into that hunger, she could fuel herself to train harder than her competitors. But in accomplishing her goals, something changed—she had sated part of this hunger. So when Barty thought about preparing for the next tournament, she experienced a *prediction error:* tennis didn't feel the way it should anymore.

Rather than ignoring her loss of enthusiasm or trying to manufacture new motivation, Barty accepted that her feelings had changed. Confident in her *perceptual inference* that competition had lost its motivating force, she made the *active inference* that it no longer made sense to train at the incredibly demanding level required to repeat accomplishments like winning Wimbledon and the Australian Open.

What makes Barty incredible is that she could listen to this feeling despite all the external pressures and the knowledge that, even if she were to scale back her training, she could likely continue to amass a fortune in prize money and endorsements. But here's the thing. Elite athletes must by definition find the drive to push harder than the thousands of others trying to climb the same mountain. Yet from Barty's account, she'd never been the kind of tennis prodigy pushed forward by an overbearing parent or coach. It's perhaps because of her inner direction that Barty is often described as "authentic" rather than just "talented" or "gifted." For Barty to remain authentic to herself, she had to listen to the change in her body and in her emotional outlook that told her she'd had enough. Her actions made her a rarity in a world in which the inability to respond to novel stimuli with perceptual inference leaves millions of people living what Henry David Thoreau called "lives of quiet desperation."

The problem with quiet desperation is that one day it all becomes too much to handle. Sonia Bansal of the Maryland

Psychiatric Research Center was interested in extreme cases of being unable to adjust to incongruities between perception and reality. She and her colleagues began to explore why patients with schizophrenia persist in rigid or distorted beliefs even when presented with evidence that contradicts their mental models. She asked two groups of participants, one with and one without schizophrenia, to report the direction of swirling dots on a computer screen. In 50 percent of the trials, there was a ninety-degree change in the dots' motion halfway through the trial. Individuals with schizophrenia tended to report the initial motion and miss the change more often than healthy participants. The stronger their delusions and hallucinations, the more this tendency persisted.

Dr. Bansal's research shows that beliefs, which we assume are very different from sensations, operate using the same principles. If we can't let in new information from the world, we continue to rely on outdated views that leave us feeling we can't make use of change when it happens—we end up feeling increasingly alienated from a world that persists in changing around us.

WHAT IF YOU'RE NOT A SULLY OR A BARTY?

When we look more closely at the most challenging moments in the lives of Captain Sully and Ash Barty, one common ingredient stands out: They allowed sensation to play a significant role before they proceeded to act on the world. Sully *allowed* the unwanted stress to linger so he could focus on landing the plane. Barty *accepted* that her drive to win had waned. We would argue that behind the scenes, this moment of receptivity changed their mental models and let them do surprising things. But what about the rest of us? We all must deal with moments, perhaps less dramatic, but equally compelling, when we realize that it's our perception of the world, not the world itself, that needs updating.

Imagine you're in line to check out at the grocery store on a busy Friday afternoon. Having placed your items on the conveyor belt, you fumble for your credit card, only to realize you don't have it. In this moment, your stomach begins to churn, your plans for dinner are blown, and all the while the cashier is staring at you, wanting to know "Cash or credit?" The people behind you in line have begun to fidget, then share conspiratorial glances. And then what happens? You have a panic attack? You castigate yourself for being an idiot? You try to find a scapegoat for the missing card ("My kids distracted me as I was going out the door")? And most important, your whole weekend looks to be ruined as you're doomed to relive this unacceptable and embarrassing moment again and again.

In this scenario, active inference is simply not an option. Barring some miraculous act of charity, there's no way you're going to be able to pay for the groceries, so racking your brain for a simple fix is not going to help. But perceptual inference can help enormously. Think of the relief that would come if you *could* say to yourself, simply, "Oh, I forgot my card. How embarrassing. I guess I'll have to go home and get it." When it's obvious you can't do the expected and pay for your food, holding on to your responsibility to pay somehow isn't helpful; it would be far better to consider your responsibility to those in line behind you and find a way to park your purchase for now. After all, looking a little foolish in front of the other shoppers is hardly the end of the world. You might apologize for the inconvenience and ask the cashier to put your cart to the side until you can return. The cashier might even normalize the event by telling you it happens all the time.

The magic moment is not miraculously finding your card. The moment comes when you stop wishing for the impossible and accept that the card is not going to appear. Suddenly, your world is transformed. You can get back into action, go home, and with your stress level at a 3 instead of a 10, find the card or, if need be, take action to cancel it and order a replacement. All it takes is an update to a mental model that says screwing up is never acceptable, at least not for you. "To err is human," a fact of life that can take a huge burden off our shoulders if we simply learn to accept it. That's why *Sesame Street*, concerned about the emotional development of children, featured Big Bird singing "Oh, everyone makes mistakes, oh yes they do" again and again. We know this intellectually, but accepting our own mistakes and moving on takes effort; it takes perceptual inference.

LESSONS FROM THE EDGE

Decades of research on mental health disorders confirm the toxicity of overly rigid mental models, the kind reinforced by the DMN. Yet even for people who engage in the most extreme expressions of emotional distress—self-harm—perceptual inference lies at the heart of effective treatments. These are individuals with strong predictive models based on prior experience of trauma, abuse, or emotional manipulation that has caused them considerable suffering. Self-harm can take numerous forms, including cutting oneself with a razor, burning oneself with cigarettes or matches, punching oneself, and, in its most extreme form, committing suicide. Self-harm has, until recently, been dimly understood. People who self-harm acknowledge that while these behaviors may seem extreme, they nevertheless serve a purpose. Converting their emotional pain into physical pain provides traumatized individuals with a means of expression and relief, even though it carries considerable cost.

Marsha Linehan, an eminent clinical psychologist at the University of Washington and the founder of dialectical behavior therapy (DBT), had engaged in self-harm as a teenager. In 1961, when she was seventeen years old, Linehan was hospitalized for social withdrawal and self-injurious behavior that included banging her head against a wall. While in the hospital, she received an arduous regimen of psychiatric medications, hours of psychoanalysis, and thirty sessions of electroconvulsive therapy, but in the end, nothing much seemed to help. When she was discharged two years later, the one thing Linehan came away with was the conviction that what she needed was treatment that responded to her unique learning history. She wanted to map out which specific emotion led to which specific thought that led to the latest act of self-injury. Only by understanding the mental model she was working under would it be possible to break the chain and learn new behaviors for dealing with the raging inner turmoil. But how do you make sense of a mind during moments like these? It's like decoding a ship's distress signals in the middle of an ocean gale.

Linehan's singular insight was that none of this could be fixed without first stopping the attempt to fix things, which means accepting turmoil as the starting point. She called this acknowledgment of life as it is, not how one wants it to be, *radical acceptance.* Perhaps this sounds like something you'd find in a Zen monastery or a crystals shop, but here's the thing: radical acceptance doesn't require chimes or incense. It's inherent to the predictive coding models of the brain, baked into accounts of how children learn about the world. The problem is that we've disparaged the role of acceptance in adult life.

The radical acceptance skills taught in DBT, like perceptual inference, emphasize sensory saturation and exploration as a first step in responding to the pain caused by injurious feelings and beliefs. Skills such as willingness and distress tolerance teach people how to purposely connect with how feelings like anger, impulsivity, and shame feel in the body, even when it's incredibly uncomfortable. Only then can we improve our relationship with these mental states and respond in ways that are not dictated solely by prior defensive habits.

What makes Linehan's work so compelling is that she was able to improve the lives of people who self-harm at a time when few effective treatments were available. She put the treatment of self-harm and borderline personality disorder on the map by providing a means of care—DBT—and by establishing its evidence base.

In the 1980s and '90s, Linehan conducted randomized controlled trials that rigorously evaluated DBT, often against other active treatments. In one study, for example, she compared outcomes for people who received DBT with outcomes for those who were treated by self-identified experts in the community. People who received DBT were hospitalized less frequently, spent less time in treatment, and made fewer suicide attempts. The deceptively simple act of radical acceptance created a pathway toward healing despite the therapeutic community's commitment to active inference strategies designed to "rewrite" inaccurate thoughts. In doing so, Linehan gave hope to countless people for whom active inference strategies were not helpful.

Regardless of whether we have ever felt overwhelmed enough to consider self-harm, our brains work the same way. We all experience situations in which we can neither walk away from nor change our circumstances. The perceptual inference practices found inside DBT are just as applicable, even if the magnitude of suffering varies.

Exercise: Finding the Feeling Tone in Sensations

As DBT has demonstrated, becoming more flexible requires being able to question the accuracy of our mental models. In other words, we must prepare for perceptual inference rather than leaping into corrective action. This means that when we experience intense feelings, we must learn to explore them instead of immediately acting to resolve them. So, when you feel angry or humiliated because you think you're not as good as others, don't push these mental states away. Instead, focus attention on where you feel the distress in your body.

In chapter 3, we went beyond our expectations and labels and monitored the ebb and flow of sensations as they occurred. Now we suggest exploring the subtle feeling tones that accompany each sensation. A feeling tone is not as much a label or an idea about the sensation as it is an inclination to do one of three things with it: it can orient you to *want more* of a sensation (sinking into a warm bath or a first bite of chocolate, the opening chords to your favorite song), *want less* of it (touching a hot stove, biting into rancid fruit, fire alarm), or just feel *indifferent* to it (pulling on a sweater, drinking from a water bottle, seeing a billboard on a highway).

Just knowing that our sensations have associated feeling tones doesn't help us much. But in practice, labeling moments of "I like this," "I don't like this," or "meh" that arise with a sensation can lessen the need to respond.

Paying attention to visceral sensations may seem trivial—after all, aren't you already aware of your feelings? Possibly more than you would like to be? But updating ourselves on how we respond to sensation is an act of perceptual inference. Recognizing feelings, even upset or aversion, separates the feeling from the action that usually accompanies it. The first step toward freeing our feelings from the House of Habit is to give them somewhere else to live.

1. **Feeling Tones in Seeing.** Find a wide-open space to look at. (If you're indoors, look out a window.) Take a moment to explore the whole scene. Now select one object in your field of vision—it could be near or far—and start to focus on the visual elements that stand out. For example, if you're focused on a building, notice parts that are darker and lighter, shadow and sunlight, smooth and textured, stone, glass, height, width. Then evaluate your subjective response. Do you like or dislike any of these features? Do you remain indifferent? Does your reaction make you more or less curious about continuing to look?
2. **Feeling Tones in Touching.** As in chapter 1, ask a friend to collect a variety of small objects for you and place them in a bag. They can be anything—rocks, a pencil, a tennis ball, shells, leaves, or rice. Now place your hand or hands in the bag and feel each item, touching it and exploring it. Can you detect a subtle sense of liking, disliking, or not caring about any of the textures or surfaces you feel? Do these reactions leave you more or less curious about continuing to touch the object?
3. **Feeling Tones in Hearing.** Whether you're indoors or out, make a point of stopping for a minute to listen to the

sounds around you. But rather than searching for anything in particular, just let the sounds come as they will. If no sound is present, pay attention to that. Can you detect a subtle sense of liking, disliking, or not caring about any of the sounds you hear? Do these reactions make you more or less curious about continuing to listen?

4. **Feeling Tones in Smelling.** Let's return to finding five things in your house or outside, some familiar and some unfamiliar, that you can explore through smell. Starting points might be coffee grounds, a bath towel, cooking spices like cinnamon, ginger, or curry powder, a bar of soap, boiling water, potting soil, your underarm, wet leaves. This time, as you inhale, can you detect a subtle sense of liking, disliking, or not caring about any of these scents? Does your reaction make you more or less curious about continuing to explore smell?
5. **Feeling Tones in Tasting.** The next time you start to eat, see if you can register the tastes that are present in your mouth. Tastes may be easier to isolate at the beginning of a meal than later on, as they mix with other food. As you chew and taste, can you detect a subtle sense of liking, disliking, or not caring about any of these sensations? Do these reactions make you more or less curious about continuing to explore taste?

Through the exercise described, you might not have changed any of your opinions about the things you saw, touched, or tasted, but that's not the objective. The goal of the exercise is to move you beyond confirming predictable expectations and instead to amplify the raw experience in front of you.

It can be difficult to stay in the moment and avoid reacting to what you find. With some practice, though, it's possible to experience perceptual inference, even in a world that pressures you to react, regulate, and respond.

There's actually a name for these magic moments. It's called being present. The feeling of *presence* is engaging with the sensed reality—full stop. You're expecting nothing more and nothing less. Being present removes the discrepancy between *is* and *ought.* And just as we find agency in creating effective change in the world, we experience presence when we learn to let go of our misfiring, habitual mental models and build predictions that better align with the world as it unfolds.

6

THE POWER OF SENSE FORAGING

I wandered lonely as a cloud
That floats on high o'er vales and hills,
When all at once I saw a crowd,
A host, of golden daffodils

—William Wordsworth

Archimedes had a problem. Hieron II, the king of Syracuse, suspected that his new crown was not pure gold, so he charged Archimedes with finding the truth. The weight of the crown could be compared to the weight of pure gold using a scale, but that was no guarantee that it was gold all the way through. Archimedes needed a way to confirm that the crown took up the same amount of space as its weight in gold. But how? He was *stuck* with the tools of the day, locked into a riddle he couldn't solve. Clearly, he needed a break, so he slipped into a warm bath. It was as he felt the water rising to match the weight of his body that his mental logjam broke apart. "Eureka!" he shouted (and, some say, went running out into the street naked). He had discovered displacement as a way of measuring mass, not simply by bearing down with his formidable powers of analysis but by letting go of cognitive striving and surrendering to sensation.

In 1666, with the plague raging through England, Sir Isaac Newton left Cambridge for Woolsthorpe Manor, the Lincolnshire sheep farm where he was born. Forced to set aside the intensity of his studies in mathematics, he found himself with unaccustomed leisure, sitting under an apple tree, in a self-described "contemplative mood." Then he saw an apple drop to the ground. In a sudden flash, he realized how incredible it is that objects don't fall sideways or upward but only downward. Somehow, these objects must be attracted to the mass of the earth. To make sense of this observation, Newton posited that some invisible force called gravity must exist. In that moment of truly attending to his senses, Newton saw the familiar act of objects falling as if for the first time. Breaking free from his driven and purely intellectual habits helped him revolutionize our understanding of physics.

Genius requires doing something new. It has many mysterious ingredients, but among them is the ability to break free from looking at things in the same old way. Many great thinkers throughout history began their discoveries by first realizing that they were *stuck* in the routine and the familiar. Yet their stories of astonishing insight rarely valorize grinding away at a problem until an answer emerges. Instead, we see that the grind is rarely where discovery occurs; the breakthroughs often come when the thinker sets aside their customary thought process. The interruption came not from another thought but from a *sensory* experience that helped them escape their DMN's House of Habit. The mental intensity subsides, and in that moment of receptivity and presence, the penny drops, just like Newton's apple.

INSIGHT IN THE BRAIN: A MATTER OF SALIENCE

We can't rewind history and do brain scans on Archimedes and Newton, but we can use our existing framework to characterize the dynamics that must have been at play in their eureka moments. The default mode network is actively rehearsing and replaying the problem at hand as well as past solutions, providing a context for active inference. The dorsal attention network is guiding the process of reflection and analysis. And then something happens that isn't covered by either of these two networks. Amid the cacophony of cognitive activity, the unexpected suddenly dominates the mental landscape. The *weight of water displaced* in the bathtub, the gestalt of the apple being *pulled* toward the earth, captures the great thinker's attention.

In this magical moment when cognition stops striving and sensation floods in, disrupting the DMN, all the habitual patterns and strategies are diminished if not displaced. What becomes *salient* is no longer the familiar, intellectual attempt to analyze one's way out of the problem. Instead, the disruption of DMN dominance allows for new representations, a network responding to sensory inputs, as per Archimedes's bath and Newton's garden landscape. Archimedes's bathwater is recast as a measurement tool; the plunk of an apple falling from a tree becomes a sign to Newton that a fundamental principle of nature had hitherto been overlooked.

The good news is that the capacity to disrupt the DMN to gain new insights is not reserved for geniuses attempting major advances in physics. Evolution shaped our brains with a system for disrupting ongoing activity whenever something shows up that absolutely demands our attention: a tiger emerges from the high grass, your boss stares daggers at you in the morning, you feel a sudden shortness of breath for no particular reason. This is the *salience network*, and it reallocates neural fuel—sugar and water—from brain regions that conduct business as usual (DMN) to regions that take in new sense data. The salience network cues us that things are not proceeding as expected—there's been *a prediction error.*

We can then decide to exploit existing knowledge supplied by the DMN, leveraging the dorsal attention network to tweak our preexisting plans—this is *active inference.* But the salience network's alarm bell can also be a cue to disrupt routine, take in new information from sensing regions in the back of the brain, and view the situation anew via *perceptual inference.* Because the sensory regions occupy different neural real estate and operate independently from regions responsible for thought and action, the salience network has the ability to shut down the habit system in favor of urgent updating. You can think of sensation as your built-in DMN ejector seat.

Getting up, stumbling into the bathroom, and brushing your teeth while half-asleep is a classic default-mode-network operation. But imagine that, halfway through your preparations, you realize that the tube of white stuff you're squeezing onto your toothbrush is not toothpaste but the antifungal cream you just bought for athlete's foot. Time to wake up and regroup.

The power of the salience network was powerfully revealed in research done by William Seeley, a neurologist and pathologist at the University of California San Francisco. He studied a relatively uncommon disease known as frontotemporal dementia (FTD), which targets the brain's salience network. Unlike Alzheimer's disease, in which memory is one of the first things to go, in FTD, people remember things just fine—they simply can't control their habits and impulses. A person with FTD could tell you that brushing with antifungal cream is wrong but end up doing it anyway. The profile of FTD teaches us that the salience network is needed to mediate the relationship between one's habits and the world. When the salience network is impaired, new knowledge can't disrupt habit, and the intellect can't be roused to help.

We all have our blind spots and habits we know intellectually aren't working but just can't shake. But unlike that of a person with FTD, our salience networks are (mostly) intact; embracing sensation allows not just a slight movement in preferences but a radical shift in the script. We're not just repainting a room in the House of Habit; we're doing a complete renovation.

The escape from conventional perspectives has long been the calling card not just of brilliant scientists but also of serious artists. Underground poet Charles Bukowski chronicled life on the margins, including his own long-standing addiction to alcohol. He believed that he couldn't write without drinking and that being devoted to writing meant embracing a habit that would destroy his health. "Find what you love and let it kill you," his commonly (mis)attributed motto, represents his ethos pretty well (whether or not he actually said it). For Bukowski, the greater danger came from conformity, a life dedicated to the rule-following habits of the DMN. Sure, getting intoxicated to create was yet another rule, but this rule successfully fed his artistic creativity—foraging for ways to defy convention worked for Bukowski, at least from the perspective of being an artist (his liver may have disagreed).

What if Bukowski is at least half-right? What if sticking to what we know is blocking the path toward greater creativity, more fulfillment, and a happier life? What Bukowski was interested in killing was knee-jerk reliance on the known, predictive models.

It's only when we go beyond exploiting what we know that we can find the freedom to create something new. Just as Newton and Archimedes discovered, the first step toward insight is stepping away from the House of Habit. And happily, it doesn't require abusing alcohol or drugs, or any other form of extreme behavior.

YOUR ALARM RINGING IS NOT THE SAME AS WAKING UP

We need a way of disrupting the DMN that isn't destructive and that isn't just for emergencies. Our disruptor needs to be both easily accessible and sustainable over the long haul. The trouble is, the salience network can interrupt the default mode only temporarily, and the window of opportunity for change is brief. All too quickly the alarm bell fades and the comforts of habits resume their appeal.

What makes a response to a prediction error sustainable is *resolution* of that error. If we simply distract ourselves from the problem, the underlying motivation to deal with the error will remain. As a type of alarm bell, the salience network does not solve problems. It simply flags some aspect of our experience for review, alerting us that a prediction error has been made. It's like someone honking as they pass on the highway to alert you that your taillights are out. To respond skillfully and make use of the warning (the *active inference* of checking your taillights), you must first see the other driver's honking as potentially helpful and benevolent. Yet if you are a jaded urban driver, your model may be that most other motorists are selfish and do not have your best interests at heart. From this model, the active inference to check your taillights is not available. Instead, you might decide that some drivers are jerks and wave them on. But that does nothing to resolve the issue. Eventually maybe a stronger error signal will occur—flashing lights of the police behind you, ready to cite you for driving with no taillights. Eventually, the prediction error becomes large enough to force you to accept an ignored reality.

But you didn't really need to get that ticket. In the face of the first prediction error—the well-meaning motorist trying to tip you off—it is also possible to engage in perceptual inference, to try to explore and understand why the driver might be signaling to you. Rather than working from a dominant model that honking drivers are jerks, and that you keep your car in tip-top shape, perceptual inference might suggest that you reexamine your assumptions.

What stops us from benefiting from informative surprises is the inclination to treat the salience network as an alarm to be quickly silenced. We are conditioned to respond to this alarm exclusively with active inference, treating the network's activation as the problem to be solved. This attitude is understandable in a demanding and often overwhelming world, but it also limits our ability to learn in favor of familiar strategies. The feeling of surprise can be an invitation to discovery and flexibility in how to respond; it is a pity that most of the time we resent the salience network's alarm signal and respond by doubling down on well-worn strategies that prioritize being right over discovering something new.

SENSE FORAGING: A TOOL FOR GETTING UNSTUCK

Getting a ticket sucks, but it's hardly the end of the world. The stakes are much higher, though, when instead of a taillight being out, there's something off in how we see ourselves and the world. Distorted mental models reasserting themselves is especially problematic in the case of depression. We know that these models have power because even after recovery, formerly depressed people have a much higher rate of becoming depressed in the future. Yet not everyone who recovers from depression succumbs again. It is as though some people are able to update their models of self, world, and future and others are not—a classic case for perceptual inference. We set out to examine whether perceptual inference determines why some people relapse after recovery while others remain depression-free.

We enlisted 166 people who'd received one of two treatments for preventing depression's return: cognitive therapy (CT), in which you learn to talk back to your negative thoughts, or mindfulness-based cognitive therapy (MBCT), in which you learn how to meditate.[1] All had attended eight weekly group classes focused on helping them minimize the effects of rumination, worry, and negative emotions in their lives. We followed these people over the course of two years as they reported back on what they were doing to stay well. Regular clinical monitoring showed that thirty-six participants—22 percent of our sample—suffered a new episode of depression. This figure was about 10–15 percent lower than that found in studies of people recovering from depression who did not do any preventative therapy, and so it seems that our folks must have been doing something right!

But what, exactly, was the "right" thing they were doing? Had the CT group become experts at negotiating with their thoughts and were they now living lives of contentment and satisfaction? Had the MBCT group learned to sink deeply into states of inner peace?

Our participants kept us up to date on how they fared on thirty-four variables previously linked to preventing depression.

These included stress, rumination, worry, body awareness, optimism, emotional intelligence, life satisfaction, self-compassion, and mindfulness. If you do the math, 34 questionnaires × 20 items each × 8 administrations × 166 participants contributed nearly a million data points toward answering the simple question, What exactly are you doing to stay well?

To shrink this massive data set into a format that was interpretable, we identified common themes underneath the sea of questionnaire responses we received. This kind of *factor analysis* was like measuring the ocean's temperature by tapping into its deep-flowing currents rather than measuring at the water's wavy surface.

We found that the greatest value derived from either form of therapy was not specific to meditation or reappraisal strategies emphasized by MBCT and CT, as their branding would suggest.

Wait—what? Mindfulness sounds like sense foraging, as you are spending a lot of time looking at the constituent elements of body or breath sensation—learning to break up a solid mass of experience into bite-size chunks that can be explored. But cognitive therapy doesn't involve foraging in body or breath sensation. And you *could* do CT by just trying to overwrite problematic thoughts and beliefs without much exploration.

What we realized was that the CT group still provided elements of sense foraging. While historically, CT attacked problematic thoughts and beliefs, the more modern form of CT used in our study focused on creating a space of inquiry in which one doesn't know how things are going to play out. Experienced cognitive therapists call this "guiding discovery" as opposed to "changing someone's mind."[2] The practices included in CT now involve ideas such as "behavioral activation" and "savoring," the idea that if you put yourself into strange or ambiguous situations, where you aren't 100 percent sure the old depression script will apply, how you see the world can open up in unexpected ways. This is a far cry from just undermining a person's negative beliefs about self, world, and the future.

The take-home message of the study was that effective therapy gets people to engage in an exploratory act, regardless of how the therapy is branded. It creates a safe space where a client can put themselves in a situation where they honestly don't know how a new behavior will play out. In the face of not knowing, all there is to do is be receptive, to explore the consequences of this novel tack. We called this *sense foraging*. The data suggested that sense foraging consists of two complementary abilities: *decentering* and *distress tolerance*.

Decentering means stepping back to observe one's thoughts, feelings, or sensations without quickly reacting to them. This approach allows us to distance ourselves from our own minds so that we can make informed and rational decisions to deal with whatever happens to be flashing through our minds.

Distress tolerance is the ability to bear the unpleasantness of what we might find. It's like continuing to watch a scary movie even as the tension mounts.

Over the two years of follow-up, participants who managed to stay well and avoid becoming depressed again were those who showed impressive growth in their ability to practice *decentering* and *distress tolerance*, the sense-foraging skills they learned in therapy. Those who relapsed into depression showed only limited growth in these skills or, in some cases, actual decline.

Sense foraging was the common mechanism in both approaches. This was both interesting and challenging to reconcile. Protection from depression wasn't limited to just one way of foraging, whether attending to the breath during meditation or journaling life experiences in CT. The benefit was shared by anyone empowered to engage in sense foraging, to explore who they had come to be in the world.

One of the participants' most frequently reported insights was that "thoughts are not facts." This is a powerful indication that a person is experiencing decentering—relating to their thoughts and beliefs in much the same way that you might relate to a physical itch or a twitch. Meditation lets you become familiar with sensation, then invites you to view your sometimes distorted thinking not as the dominant narrative but as just another sensory event. CT invites you to loosen your view of thoughts as facts and approach them as ideas or interpretations that may or may not be true. In both approaches, participants found a way to question the truth behind even well-rehearsed mental chatter.

Rather than championing one therapy over another, the data showed that what mattered was the ability to practice sense foraging, to be curious about prediction errors from the world, and to tolerate the ambiguity of what might be causing them. We became intrigued by the idea that there might be additional ways for people to increase their sense-foraging skills.

RESILIENT PEOPLE HAVE SENSING BRAINS

In addition to filling out questionnaires before and after preventative therapy, we took advantage of our two-year follow-up design to do a little brain scanning. We had participants watch sad and neutral film clips in the scanner, giving us a snapshot of their emotional reactions. Then we waited two years to analyze the data, until we knew who had relapsed and who had stayed well. In fact, this became the largest neuroimaging study of depression following preventative psychotherapy ever undertaken.[3] The purpose was to meld a rigorous analysis of what participants experienced—the decentering and distress tolerance and all—with an account of how the brain changed to support well-being.

Despite some participants' reference to other factors such as curiosity, perspective taking, and tolerance, the brain data convinced us that sensation is really where the action is. We knew from our earlier work that the sad films would likely activate the DMN at the expense of sensory processing, but we weren't sure if that would affect depression vulnerability over time.

Despite the power of the DMN, the loss of sensation was the key. Just as before, sadness shut down sensory processing. Some of our participants had retained niggling symptoms—a feeling of guilt here, a lack of motivation there—and the strength of those residual symptoms was positively correlated with the extent to which they'd shut down. In addition, some participants had experienced a greater number of past episodes of depression, and we found a similar association between this frequency and greater sensory shutdown. It appears that the triumvirate of feeling sad now, feeling depressed more generally, and having had more depression in one's past all converged, and they showed up in the somatosensory cortex, which processes the feelings from the surface of the body, and in the insula, which processes the body's internal state. A higher-than-average intensity of somatosensory shutdown made a person *twenty-five times* more likely to relapse than someone who only pumped the sensory brakes instead of slamming on them.

This doesn't mean that there was no hope for these folks. Therapy didn't reverse the pattern of shutting down, but we found a second signal, in the front of the brain, responsible for planning action rather than sensation. This area tended to quiet down after the individual worked on mindfulness or cognitive therapy techniques for eight weeks. If this region quieted down more than average, it cut the risk of relapse by more than half.

We examined what this frontal region talked to in the brain and found that its activation was linked to suppression of sensory regions. Using this region to react to stress with thinking, planning, and fixing was like pumping the brakes on sensory regions. While eight weeks of therapy wasn't enough to stop shutting out sensation, taking one's foot off the brakes was a good start.

The questionnaire data told us that participants were relying less on knee-jerk reactions to sadness. They were learning to stop trying to suppress or control emotional experience by broadening their palette of acceptable experiences and learning to go along for the ride. After eight weeks of training, their brains were still hardwired to avoid sensation, but those who got something from therapy learned to stop riding the brakes at the first sign of trouble.

We'd discovered something amazing. For the first time, we showed that inhibiting sensation in the face of stress was a brain marker of one's past, present, and even future depression vulnerability. Previously we had shown that sensory inhibition was linked to depression in the moment; now we knew that blocking out sensation predicted future bouts of illness. These findings were totally different from the prevailing wisdom that blamed depression on too much activity in the front of the brain. Yes, the front of the brain is important, but vulnerability occurs when the front of the brain squeezes out new information from our senses.

In other words, we had discovered the strongest evidence yet that the momentary act of sensory inhibition powerfully affects how your life plays out. On the bright side, our resilient participants told us how they had learned to cope with this pattern—it takes a growing willingness to forage within unpredictable and disruptive sensation. In the face of stress, it takes courage to look and to tolerate what one finds.

Exercise: Foraging at the Grocery Store

We suspect that there are situations in your life that are recurring and predictably unsatisfactory. You know you tend to do *x*, and it leads to *y*, and you don't like *y*. What is something else you can do (not just "not *x*," but really, something *else* you can do)? Look for actions where you honestly don't know how they will play out. Recalling Rumi's invitation to play in the face of discomfort, what does your musical instrument look like, an act distinct from resigned study and reading?

As a concrete example, you could try going to a larger grocery store. The produce section contains a vast array of options, but at the same time, you already know what you like. From the dozens of different types of fruits and vegetables, there are probably just a few that you tend to buy, and maybe a few other "sometimes" foods, and the rest, your eyes just slide over.

This time, take a notebook with you. Your task is to record ten different colors you can spot among the produce. Then list five different smells you can pick out. Finally, make note of any other sensory information that grabs your attention as you meander through the aisles. That's right, meander, don't rush; see if you can be receptive in the face of not knowing what will capture your attention.

Now, if you were to make your fruit or vegetable purchases for the week, what would they be? You might get what you always get, and that's fine! We wonder whether you might have become more curious about flavors that you would typically ignore or avoid. The goal here isn't to stay the same, nor is it to force you to pick a food you hate. It is about becoming curious about your choices, seeing how this immersive sensory environment can speak to you, making it more than just another trip to the store.

7

PRINCIPLES OF SENSE FORAGING

The real voyage of discovery consists not of seeking new landscapes but in having new eyes.

—MARCEL PROUST

LEARNING TO FORAGE

René Redzepi, owner of the renowned Copenhagen restaurant Noma, turned to sensation because he was being crushed by success. As his place became more and more popular, the pressures mounted, and René found himself engaging in the same bullying behaviors his mentors had tormented him with when he was a chef in training. He'd sworn he'd never follow that path, so he did something radical: he took a break, temporarily closed his kitchen, and headed into the woods.

Determined to explore the art of food foraging, Redzepi was shocked to discover that a single blade of grass could taste like cilantro, and that an ant (an *ant*!) could taste like lemon. He was able to find all these new and unexpected flavors just a few steps beyond his doorway. But even more unexpectedly, Redzepi's relationship with food changed, and then his relationship with the entire world changed. He began to realize that being a chef was not just about making raw ingredients delicious through culinary alchemy; he began to see himself as a part of nature, with a responsibility to preserve and protect it. He began with a hunger for exciting new flavors, only to realize that what he really craved was this sense of connection.

Competitors and critics scoffed, but René demonstrated that by shutting down business as usual and engaging in this kind of sensory exploration, not only could he enjoy collecting local and wild food, but he could fashion it into world-class cuisine as well. Tirelessly combing the Nordic coastline and forests, he brought back edible flowers, sea buckthorn shrubs, kelp, shellfish, berries from rosebushes, wild mushrooms, garlic, and root vegetables to feature on the menu. Redzepi showed that with the right visual, tactile, and olfactory sense training, it's possible, as he put it, to "stroll through the wild and pluck things like we do from the shelves of the supermarket." Unique items like snail broth and kelp ice cream helped Noma earn the title Best Restaurant in the World four times in a period of ten years. But just as important, this break from established patterns to freely explore the sensory world helped nourish Redzepi and his staff. It made the work they loved (but had found increasingly stressful) new and exciting again.

No matter what kind of career we have, we all have parts of our lives where innovation and discovery would be helpful. Like Redzepi, we may be run down by the monotony of the everyday, feeling burned out by the same routine. The obvious question is, Why keep doing the same thing? Isn't there a middle ground between more of the same and walking away from your responsibilities? The answer is unequivocally yes. Rather than becoming resigned or avoidant, both of which close us down to new possibilities, you could try your own version of Redzepi's combing through the woods and along the shoreline. You could try *sense foraging*.

Imagine what might be possible if our minds were a little less rooted in habit and our behavioral repertoire were expanded. Just imagine what might happen if there was a new way to do your job, talk to your partner, or feel about yourself. What if you're stuck only because you've stopped exploring, which has left you drained of any fresh ideas? That's what sense foraging helps you do—explore.

It might turn out that your job really does stink, you and your partner really do have irreconcilable differences, and you really are all those nasty things you've called yourself. But don't you owe it to yourself to find out if there's more to your malaise than *everything sucks and it's all my fault*?

Even if exploration doesn't help you find a solution to your most deeply entrenched problems, you'll be better off for having looked. A recent study in the *Journal of Happiness Studies* found that, on stressful days, people who managed to remain curious felt better than those who didn't.[1] Their openness to exploration meant that they saw their challenges as opportunities to learn and grow. Again, we see the contrast between the false refuge of the House of Habit and a willingness to leave home and explore.

GETTING STARTED

Despite his success, Redzepi needed to set aside his familiar routines and well-established knowledge to branch out and rediscover his passion for cuisine. But whether we're on top of the mountain as he was, just starting out, or still struggling uphill, each of us can benefit from exploring our own wilderness of new experiences, insights, and ways of knowing just outside the safe confines of habit and routine.

Even so, letting go of the tried and true is scary, because you have no idea what's going to replace it. That's why far too many people wait for a crisis to force them to change. Their marriage falls apart, they get fired, or they wake up in a jail cell charged with drunk driving. Most people don't have a clear sense of just how stressful, unmanageable, or unrewarding their lives have become until after they escape the House of Habit and reach the open air. So where does the motivation to change come from before circumstances force change upon us? Change can seem so demanding that we tell ourselves we simply don't have the time or energy to think about it. But that's how we roll along in a rut, until it leads us into a very deep and very unpleasant ditch.

Happily, Redzepi didn't wait until a crisis forced his hand. He saw the signs early and had the courage to find a new way—and it wasn't as if he had a lot of time on his hands. If you've read Anthony Bourdain's book *Kitchen Confidential*, then you have some idea of just how insane a chef's life can be. Imagine Redzepi, exhausted from another seventeen-hour day, worried about cash flow, staffing, ingredients shipments, competitors, and then, on top of all that, his growing sense of meaninglessness. Serving food is an intensely competitive business, and if he'd tried to ignore the signs of burnout and simply kept on keeping on, it's not hard to imagine that the quality of his cuisine might have slipped, he might have lost it with some customers and driven them away, and staff might have left him for a less crazed work environment.

Sometimes the numbing effects of habit can conceal just how bad we feel, and the blinders imposed by the DMN make it hard to imagine just how different life could be. Which is why it's so important to latch onto that first moment when *you notice you're stuck*. That's the first step in getting unstuck.

STEP 1: FIND YOURSELF IN A STICKY SITUATION

Once we see that there really is a reason for change, a difference between the way things are and how they ought to be, we can find the motivation to take the plunge.

In the world of championship poker, not acting as your rational, best self is known as being "on tilt," like a pinball machine skewed from its level position. Children are "on tilt" when they're petulant or throwing a tantrum. Even as an adult, you can be "on tilt" when your blood sugar's too low. You become hangry, and suddenly your coworkers are jerks, your partner is annoying, and the smallest demand can feel like the last straw. Our colleagues, friends, or partner may not know that all we need is something to eat, but they're well aware that we're irritable and cranky. The problem is, we lack awareness that our internal state has changed, so we blame everything that's annoying us on others. Then—sheepishly—post nap or snack, we realize that everyone around us isn't impossible and that we may have been overreacting a little.

Emerging neuroscience casts the blame for the "hangry" phenomenon on the release of a chemical in our brains called neuropeptide Y, which is related to feelings of anger and aggression.[2] Neuropeptide Y seems to lead to a more intense but unstable activation of the salience network, the system that alerts us to prediction errors, telling us that something is wrong and needs fixing.

Being hangry actually doesn't intensify the habits enshrined in the default mode network, but it puts us on high alert, which causes us to be more reactive to any violation of expectation.

Ultimately, the "correct" response is one of perceptual inference: "Things are getting intense. I wonder if it's me." But as we've seen, we're not skilled at this kind of introspection. Instead, we fall into active inference by default, seeing culprits everywhere we turn.

There's an active inference that could solve the problem—grabbing a Snickers bar (or, ideally, something that provides a gradual sugar build rather than a sugar rush). The hard part is being sufficiently self-aware to *notice* the problem—that's a critical perceptual inference.

When the culprit is our own metabolism, railing against our associates and bemoaning our lot in life accomplishes nothing. We're stuck in the House of Habit when we should be foraging in the fridge. But first, instead of lashing out, we need to look within.

STEP 2: LOOK BEFORE YOU LEAP

A reliable marker on the road to health and happiness is remaining curious about the mental habits that run our perceptions behind the scenes. The next step toward improvement is to *prepare* an appropriate place to forage. Anywhere you can explore your senses can serve for practice, but that doesn't mean all choices are equally safe, comfortable, and effective for you in this moment. At the end of chapter 6, we suggested a trip to the rich tapestry of sensation that is a grocery store produce section. We chose this example because while it might be a bit weird to document color and smell, it is a safe place to linger and examine the produce in front of you, the urban equivalent of foraging in the wild.

It may seem pointless to pick a concrete activity for sense foraging when you could just imagine yourself doing so. However, simply predicting what would happen is to confuse good intentions with proper execution. When you are starting out, you want to pick an activity where it is safe to be a bit weird and there isn't intense pressure to rely on habit. When nothing much is going on, it is relatively easy to notice if you're ruminating when you're trying to sense forage, then to correct the behavior. It's much harder to practice in a chaotic and demanding environment—although Captain Sully allowed sensation and succeeded in landing his plane, we don't need to learn in such high-stakes environments. Be kind and set yourself up for success.

That's why picking a proper sensory target is so essential when you're setting out to do sense foraging. Archimedes noticed the water in his bath, Newton tracked apples in the orchard, and Redzepi sought sustenance in the woods. None of them just stayed in their office and imagined not being stuck. You need to *spend time* sensing to escape from the mode of predicting and rehearsing existing models. Otherwise, just predicting what it would be like to sense forage means loading up another set of expectations rather than engaging in sensation. We want you to pick a destination outside the House of Habit, lest you end up just wandering from room to room. If you can do that, you will be ready for step three, to engage in sense foraging.

STEP 3: COME WITH ME IF YOU WANT TO LIVE

Sarah Connor (played by Linda Hamilton), the protagonist of the movie *Terminator 2*, had plenty of motivation to engage in sense foraging. She was also a breakthrough female character in moviemaking because she left behind her *Terminator 1* identity as a damsel in distress to become an empowered freedom fighter.

In case you missed it, the plot puts a twist on the first *Terminator*, in which Arnold Schwarzenegger played an evil killer robot sent back in time to prevent the rise of humanity's future liberator. In the sequel, Schwarzenegger's character has been reprogrammed and sent back again, this time to thwart the very mission he'd been on previously.

At one point, he's helping to rescue Sarah from a mental asylum and in his distinctive Austrian accent says, "Come with me if you want to live." In the first movie, Sarah had heard this line while being rescued *from* the Terminator. Yet she recognized that being *stuck* in habits of fearing and fleeing from the Terminator would not be helpful in her current situation, trapped in an asylum while under assault from a new model of killer robot. She had to rapidly *prepare* to update her model of the world to one in which she and the Terminator could work together to fight against a greater threat.

The Terminator's famous ultimatum reveals that to truly change yourself, you must let the world in, shattering the part of yourself that rejects your experience of reality. At some point, you must engage in the third step and actually *do it.* And Sarah *does.* Her pupils dilate, she lets go of her need to fight or flee, and she sees clearly for the first time that this dangerous figure from her past has been resurrected as a most unlikely savior. She lets go of what she knows and rewrites her narrative based on her experience in the moment. And in this moment, she is transformed from Sarah Connor, would-be victim, into the mother of the human resistance to tyranny. Not quite the same as Redzepi finding a lemon-flavored bug, but pretty good nonetheless, no?

Fortunately, most of us will not have to engage in sense foraging under threat of death by killer robots. But unfortunately, our position of relative safety also means we won't get such an iconic instruction to dive in. But the threat of living an unsatisfying life still lurks in the background. So, consider us a friendlier, less muscular pair of Arnolds in this moment. Come with us if you want to *live*! Let's take the plunge and engage.

PRINCIPLES IN ACTION: FEEL IT IN YOUR SOLE

You may never have heard of the magician Tolly Burkan, but chances are you've heard of the self-improvement mainstay he introduced in the 1970s—firewalking. Now unavoidable on the seminar circuit, firewalking involves stepping along a fifteen-to-twenty-foot path of burning coals. But this is not just some cultural artifact from the Nixon era. Firewalking is referenced in the Bible and in Roman history and is a tradition among the !Kung bushmen of the Kalahari Desert.

Burkan himself thought that firewalking was a ruse before he tried it. In an ABC News interview, he recalled, "Immediately, I thought, you know, this has the potential to knock people out of the rut of conventional thinking . . . it was remarkable and potentially life-changing."[3]

Firewalking is not magic in the sense of mind over matter. The physics behind firewalking clarifies why it can be done safely: wood embers conduct heat poorly, foot contact time can be limited to under a second, one can avoid stepping on the hottest glowing coals, and wet feet vaporize heat into a protective steam called Leidenfrost.

Yet firewalking *is* magic as an example of *mind over mind:* we convince ourselves to touch fire, even though one of the first things we are taught as children is that some things are hot, and hot things can hurt you. To firewalk, you need to be willing to radically revise deep-seated knowledge. And after a lifetime of trying not to get burned, this is easier said than done.

From our perspective, firewalking is an extreme and concentrated form of sense foraging. And to get through it unscathed, you'd need to follow the three steps we've outlined.

- **Notice you're stuck.** There's a strong prior belief that you want to challenge: you start off stuck on the idea that any exposure to fire is harmful. You're invited to question the totality of this belief, which can motivate you to see the world differently.
- **Prepare to forage.** There's a technique for walking over coals that minimizes exposure to the hottest flames. Andrew Weil, a pioneer in integrative medicine, learned this lesson the hard way on his first attempt. "I felt mentally scattered," he wrote. "The pain was near my limit. I do not think I could have taken more than one or two additional steps."[4] Yet on a subsequent occasion, he was more present: "There was no sensation of heat, it just felt crunchy . . . I could've done it all night, it was amazing."
- **Do it!** Just hearing about firewalking may make you curious, but only seeing firewalking in action—or better yet, doing it yourself—is going to change your deep-seated beliefs.

So how exactly do we employ this same receptivity and curiosity as we confront the firepits of daily life? Our own negative feelings are like that carpet of burning coals. The way through is to approach our feelings as valuable rather that seeing them as problems to be solved, avoided, or extinguished. Only when we stop avoiding or controlling will we have a chance to become more than the sum of our habits, to become something new. But to do that, we need to call on sensory experience to help us subvert the dominance of the DMN, the keeper of habit. Let's try a short exercise to get started with foraging.

Exercise: A Backward Way to Eat

As small children, we're all about sensation, even when learning how to manipulate a small stick filled with graphite and clay to write letters. Yet the moments of sensory exploration become rarer as we grow. After a while, muscle memory takes over, our hand simply assumes the correct shape around the pencil, and the letters appear on the page. Still, life has a way of pulling us back into sensation, particularly when things don't go according to plan.

Unless you've lived a charmed life, you've probably experienced an injury of some kind: turned an ankle, sprained a wrist, or broken a bone. Try to think back to that time and remember what it was like to have to learn to walk or eat according to a different set of rules. The constraints from your injury meant that your habitual movements were met with resistance, pain, or both. You may have had to change how you shift your weight when you walk, or move in bed, or manipulate utensils with your nondominant hand. In doing so, you likely had to confront the frustration and uncertainty of losing a function that was once graceful and effortless, and you noticed how clumsy the fork felt in your off hand, and how much more time it took to eat. Maybe alarm bells were going off the whole time in your default mode network as it tried to nudge you back to moving or eating as usual, while the pain or immobility reminded you again and again that, even if this were possible, it was definitely not a good idea.

Despite the challenge and frustration of having to learn how to do the basics again, wasn't there something intriguing in having to slow down and reflect on the acts of moving or eating? Were you able to spend more time actually seeing and tasting your food? Maybe for the first time you really appreciated the complexity of shifting weight and balance when walking, or even in getting out of bed. Over time, you may have experienced new confidence as you saw renewed possibilities for regaining function in the face of the injury's disruptive effects.

To get a better sense of what we're talking about, try this experiment. The next time you eat, try eating with your nondominant hand. It may be the weirdest meal you've ever had. But we invite you to explore whether this meal might feel more satisfying than usual—the feeling of exploring, adjusting, and accomplishing the basic act of self-care that feeding represents. What is it like to push back against the self-criticism of being slow and clumsy? To say yes to the embarrassment of dropping food back on your plate, of taking longer than usual? What happens to the flavor of food when you eat this way, committed to sensation and exploration and to savoring every moment? And at the end of the meal, knowing that you can go back to eating in your usual manner, you might even feel a slight sense of loss, of missing this quiet fumbling at mealtime, the escape from habit, and the discovery of a new home away from home.

If you want to go even further, consider what it would be like to eat a meal without any cutlery at all! Most of us were raised to eat with a spoon, a knife, and a fork and to appreciate that the correct use of cutlery connotes a measure of refinement. When we were toddlers, we most likely were told that eating with one's hands was wrong, even though close to a quarter of the world's population does it every day, at every meal.

Yet if we break away from the assumption that fork-and-knife culture is superior and attend to the sensations of eating with our hands, we might notice some surprising things. There's the "leave your ego at the door" feeling when you decide to pick up a handful of peas and pour them into your mouth. But something deeper might be happening, too. We may have to confront our own assumptions about "clean" and "dirty," the belief that contact between food and hands is inappropriate for a civilized adult. If that's the case, feel free to wash your hands first. Or you may just feel like it's silly and beneath you to do this—but that's the point, the "you" that is unhappy would never eat with your hands. If only just for a moment, are you willing to try out being someone different?

This exercise in sensing may not change the taste of food, but it might make you freer to experiment in the future, a baby step toward tackling the bigger problems you face. In other words, you don't have to burn down the House of Habit to free yourself from the grip of the DMN. Simply being playful with your most ingrained routines can be like cracking open a window to let in some fresh air. So, give it a try. We think you might be surprised by the person you become.

WALKING THE WALK: NINE SIMPLE RULES FOR SENSE FORAGING

If this were a book about weight lifting, this would be the section on technique and form. An Olympic weight-lifting coach knows from experience what works and what doesn't to safely build strength. Our experience working in health and wellness has helped us recognize general principles that cut across different forms of mental training for well-being. The following principles will help you recognize a good sense-foraging practice when you see it. Throughout, we'll refer back to the eating exercise to show how it checks off the boxes. Here they are:

1. **You can't force it.** Sense foraging is the art of optimizing the conditions for perceptual inference, but if you already knew exactly what to do, you wouldn't be stuck. See if you can let go of the idea that you're going to be in charge of how sense foraging turns out.
2. **You can choose it.** Happily, setting the intention to forage may be sufficient for sense foraging to occur. All it takes is that one salient moment that disrupts the default mode network and frees up resources for sensation. After that, you just have to get out of the way and see what happens.
3. **Ubiquity.** Sense foraging can be practiced anywhere, anytime, because as long as you're conscious, there's always something in your sensory world that you can attend to. It's less about finding some special stimulus and more about seeing the sensory world as always available.

   Forty years after Wilder Penfield demonstrated that mental activity could be altered by implanting a tiny electrode in the brain, neurologist Helen Mayberg applied this technique to address treatment-resistant depression.[5] For patients afflicted with this condition, whose mental habits skew toward neurovegetative shutdown, turning on the electrode reverses the effects.

The wildest part is that the electrode—and thus the effect—can be turned off and on via remote control. When the electrode is turned on, it acts like a pacemaker, regulating the flow of information along the neural circuits at the base of the front of the brain, one of the hubs of the DMN. What's startling is the experiences people report when the electrode is turned on. Although the electrode is not placed in the brain's sense factories, its effects have a strong sensory flavor, showing up as greater sensation in people who simply can't disrupt the DMN and embrace sensation on their own.

Patients report things like "I can suddenly see colors again" or that they feel "sudden calmness or lightness," "disappearance of the void," a sense of heightened awareness, increased interest, and "connectedness." One individual reported sudden "brightening of the operating room," including a description of the sharpening of visual details and intensification of colors.

This surgical treatment shows that, in even the most severely depressed, the ability to sense deeply is suppressed but not obliterated. That's good news for all of us: sensation lost can be reclaimed, even for those mired in suffering. Fortunately, most of us are not in such dire straits, and so we don't have to place ourselves on a neurosurgery waiting list to free up sensation. As we hope we've made clear, it's not at all difficult to learn to turn this switch on for yourself.

4. **Completeness.** What you sense is sufficient. You don't have to create new meanings or evaluate whether what you're sensing is good or bad. That's the slippery slope back

into expectation. Seeing what you sense clearly, with no add-ons or preservatives, is what you're looking for. Rabbit Angstrom, the confused everyman who stumbled through a series of four novels by John Updike, smacked a perfect drive on the golf course and yelled, "That's it!" He wasn't referring to how to hit a golf ball; he was referring to how it felt. A fleeting moment when everything in life seems in perfect alignment. Rabbit revels in the possibility that each moment of sensation is complete unto itself. You may have goals and expectations for what sense foraging will do for you, but you must let that all go to practice sense foraging in the moment.

5. **Concreteness.** Eventually, you'll be able to apply the techniques you've used to explore the physical world to feelings or thoughts, but the ephemeral nature of internal experiences makes them challenging targets for a beginner. Why make it hard? A good object for attention should be concrete and salient—like Rabbit's golf stroke. It should be obvious to you whether you're still paying attention to it.
6. **Immersion.** Sense foraging shouldn't feel like a Magic Eye puzzle, those abstract pictures where you see gradations of tone and try to unfocus your eyes to reveal some hidden picture. Sense foraging should be something you can ease into and feel that you've become a part of, rather than remaining an outside observer struggling to get a peek.
7. **Safety.** Your voyage into sensation should be as safe as possible. An element of danger or mystery may be exciting and help fuel your enthusiasm, but it really isn't necessary. You're already taking a risk by allowing yourself to be changed by the experience, so there's no need to take up BASE

jumping or kitesurfing. Firewalking with a skilled guide? Okay. Drunken leaping over the bonfire at your campsite? Mmmmm . . . not so much. If your sense-foraging practices seem like they might cause harm, please pause and reconsider.

8. **You own it.** Sense foraging is the art of leveraging a fundamental human capacity to explore and learn from the world. Still, you may encounter people using their credentials and expertise to gatekeep your access. They may say you have to take an expensive course or gain certification in a sense-foraging practice. And of course, expertise exists. If your sense-foraging practice is finding mushrooms in the forest, or exploring the Hindu Kush, it's well worth paying for expert guidance. But ultimately your ability to sense forage has to be negotiated by you. You are the highest-level expert in your own practice. And remember, your kitchen table can be an excellent place to start.
9. **It's awesome.** When sense foraging works, it creates a sense of presence, the feeling that the difference between your model for experience and your sensed experience has dissolved. If you get to a moment in which there is no longer any noticeable difference between expectation and sensation, a curious thing happens. The boundary between the observer and what is being observed begins to blur. Such an experience is not, generally speaking, emotionally neutral. Often, there is a feeling of the self expanding or even dissolving—it is *awesome* in the traditional sense of the word. Freed from a set of perceptual habits, we can witness the complexity and richness of the world with fear or wonder. Ordinarily we would avoid such feelings, but

in the context of sense foraging, we curate a safe space to practice, where we can learn to value these feelings as signs that we are on the path to discovery. A feeling of being surprised to see yourself as but one small part of a larger system, especially in situations you would normally take for granted, is a big green flag that your sense-foraging efforts are leading to perceptual inference.

ARE WE THERE YET?

It's easy but not simple is probably the best tip we can give you as you embark on your sense-foraging journey. Smell this, taste that, touch here, see this, is easy enough to do, but exactly how all this leads to model updating is less straightforward. When you find an access point that fits you, change will begin to occur immediately, but it will likely start small. You may feel a bit less stuck in one way of thinking and start to realize that you have this resource to go to—this place in sensation that counteracts the weight of acquired knowledge. Just knowing that you have this option can mean the world, a world less expected but present, available, and surprisingly rich.

Equally, if nothing happens for you right away or what you do find is underwhelming, it can help to remind yourself that your mental routines have been in place for much longer than your newfound desire to explore the world of sensation. The nine simple rules for sense foraging are as much about attitude as they are about the nuts and bolts of the practice. The key attitude here is patience. Don't be surprised if you must audition a couple of access points. Your first choice may not prove to be the best fit for you, maybe checking off only a few of the boxes mentioned previously.

In time you will find one that speaks to you, and, with repeated practice, you will be able to recognize when you are stuck as its own type of model—connecting the option to engage in sense foraging with the feeling of being stuck, a kind of muscle memory. But it is a special model because it is a model that points you toward exploration and updating rather than consolidating, defending, or exploiting what you already know.

As long as we need to make sense of the world, our brains will keep generating models laden with expectations. A few acts of sense foraging are not going to change this reality. But instead of being condemned to move through life from chaotic sensation to rigid judgment, there is another way the story can play out. If you incorporate sense foraging into your model for making sense of the world, there is a chance to find balance—making use of hard-earned insights while maintaining the wonder of youth.

Exercise: Just-in-Time Sense Foraging

Much of the time, transformative practices are steeped in rituals and preparation, but because sense foraging is constantly available, you can connect in the next moment, even if you have only a minute in which to forage.

Pick something you can sense right here and now and take the plunge.

You've got fingers? Great. Close your eyes—how do you know you have fingers? Can you feel them?

You've got feet? Okay. Tap 'em and feel the vibrations.

Still breathing? Are you sure? How do you know?

How's it smelling in your neck of the woods? Take one big sniff and describe what's in the air.

What can you hear? Take a minute and find out—even in a noisy environment, are there small pockets of silence?

Try to look at something around you that you normally would never pay attention to.

The point of this exercise is to show how quickly you can shift into a sensing, exploring mode. It can be helpful to know that you don't have to do a lot of planning to get into foraging. While there are benefits to establishing a deeper routine, a quick foraging break is always available when you feel like your conceptual brain is bearing down on you.

8

ACCESS POINTS

I will not have my life narrowed down.

—BELL HOOKS

FIND YOUR FIRE

Sense foraging is motivated by a feeling of being stuck or unfulfilled. When you learn to meet this feeling with sense foraging, to value flexibility over consistency, magic begins to happen. The DMN becomes more flexible despite its inherent stickiness, able to toggle out of one situation into another. You get home from work, breathe in your home life, and work just slides away, replaced by more domestic routines and meanings. You are still using your DMN, but you are letting the moment be salient and important, cuing the DMN to adapt to the present rather than pulling you into past or future concerns.

The feeling of extraneous thought dissolving away is mysterious but well-documented across different cultures and contexts. Meditation and psychedelics both describe the concept of ego death or transcendence, anesthesia and dream memories often feature out-of-body experiences, and in pop and sports culture the idea of the flow state has been popularized by authors such as positive psychologist Mihaly Csikszentmihalyi. These states are reached through different access points, but they all tap into principles of sense foraging. A DMN that truly serves you should automatically provide context for whatever it is that you attend to rather than pulling your attention elsewhere: the meditator is undistracted by thoughts of checking their phone; the objection "but this is not real" no longer limits the dreamer; the athlete simply reacts to the game around them rather than forcing a movement.

On the other hand, it can be a bit scary to let go of consistency, depending on how safe you perceive your situation and activity to be. Our DMNs weave security blankets of meaning around us to shield us from the unpredictability of the world. Sense foraging pulls at our blanket's loose threads. When done skillfully, this unraveling process frees us to move more comfortably, like tugging loose tightly tucked bedsheets. Other times, we may become disoriented or overwhelmed by the suddenly unrestricted movements of our minds. Either way, learning to recognize and value this feeling of freedom—that you have removed barriers between yourself and the world—is a signpost of sense foraging at work.

So, how do you find a safe and sustainable way to loosen up your DMN? The trick is to normalize failure. Failing to stick with a diet or to quit smoking could lead someone to conclude that the problem is too big to conquer and so resign themselves to their unwanted habit. But research shows that changing a deeply entrenched habit takes multiple attempts.[1] While you might get lucky, the average person will run into some unsuccessful access points before they land on one that works for them.

There's no way we can predict the right approach for you, specifically. What's more, you yourself may not have the same motivation or support system each time you try. The best advice we can give is to remain determined to change but not dogmatic about the method. Keep trying things that check off the boxes laid out in chapter 7, bearing in mind that most people won't find the right technique right away. This may seem discouraging, but it's not as discouraging as being sold the one "right" technique and then feeling like a failure because you're still smoking, or gorging on Oreos, or falling for every well-built guy at the gym, even if he still lives with his mother, or spending endless hours playing *Angry Birds* on your phone. We want to give you different tools with which to build your sense-foraging experience, but it is up to you to see which one fits best in your hands.

TOUCH GRASS (NATURE)

Today's kids spend a lot of time on screens, and some have learned to recognize the signs of losing touch with reality. To signal to another internet user that they might be mentally stuck in a virtual world, they recommend that this person "touch grass," which means get outside and reconnect with the real world. This idea is echoed by neurologist and bestselling author Oliver Sacks, who famously said, "In forty years of medical practice, I have found only two types of non-pharmaceutical 'therapy' to be vitally important for patients with chronic neurological diseases: music and gardens."[2]

It turns out that there's a huge amount of scientific literature on the benefits of regularly "touching grass." From the self-discovery of Australian aboriginal walkabouts to the relief that hiking and camping provide urban dwellers, spending time in natural environments is as close as we get to a timeless, universal good. Experimental studies have shown that access to green as compared to urban spaces improves attentional control, cognitive flexibility, working memory, and recall. In addition, green spaces near schools have been shown to promote children's cognitive development and self-control behaviors. Large population studies have found green space to be linked to increases in positive moods and decreases in stress hormones, blood pressure, and muscle tension.[3]

One theory suggests that living in artificial environments requires constant use of focused attention and problem-solving skills and that going out into nature is restorative because it allows us to let go of this heavy lifting and let the brain recharge.

But how much time do you need to spend in nature to start reaping some of these benefits? One answer, derived from survey data on twenty thousand people, indicates that every hour you spend in nature improves your chance of reporting good health by about 2.5 percent. These benefits continue to accrue up to around three and a half hours per week before things level off—that amounts to thirty minutes a day or a few longer excursions each week. Compared to people who spent no time in nature, people who spent three and a half hours or more had a 7 percent higher chance of feeling healthy.[4] It didn't seem to matter whether the two hundred minutes were accumulated in one long session or shorter bursts. This is good news in that it allows people to be as flexible in their sense foraging as their schedules permit.

To better understand why spending time in nature can lead to health benefits, Craig Anderson and colleagues compared the effects of different outdoor activities. They explored whether nature experiences had to be exceptional, like mountain climbing,

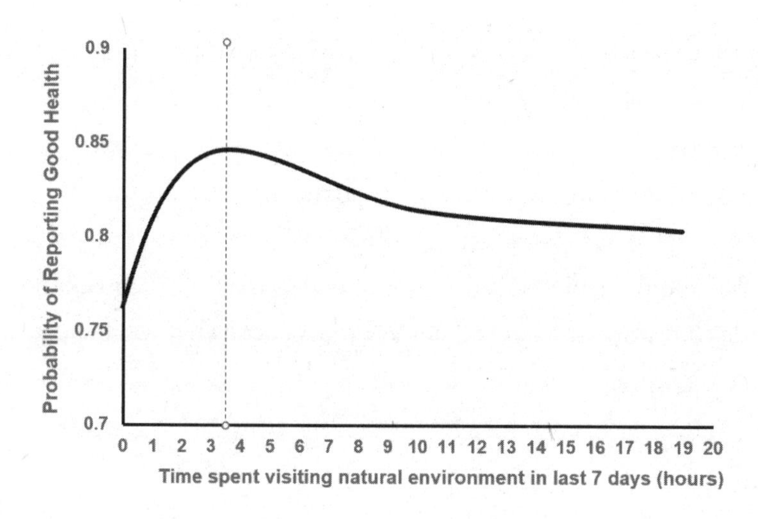

to be beneficial, or whether ordinary activities would do, like walking through a park. The answer was: it depends on how you experience the activity. The researchers found that nature's restorative power comes when the activity induces a feeling of suddenly being connected to a larger world rather than confined within your own skin—this is also known as awe.[5] The feeling can be induced by immersion in any natural environment, whether exhilarating (white-water rafting) or simply beautiful (watching a sunset). Anderson and colleagues' result held up across diverse groups of people, including at-risk youth, veterans, and college students—nature matters when it inspires us to step beyond the self.

While descending Mount Katahdin might give you the type of transcendent awe that nearly sent Henry David Thoreau over the edge ("This was that Earth of which we have heard, made out of Chaos and Old Night. . . . It was Matter, vast, terrific . . . rocks, trees, wind on our cheeks! the *solid* earth! the *actual* world!"),[6] even small moments of beauty, surprise, or enthusiasm can pull us out of our fixed way of knowing and be transformative. This is the "broaden and build" theory of positive emotions, popularized by social psychologist Barbara Fredrickson. Sense foraging is about creating the conditions where we can notice these typically mundane positive moments and get the ball rolling.

In 1993, the first issue of *Wired* magazine started using the words "tired" and "wired" as a social barometer of people and trends that were considered passé versus those that were ascendant. We've co-opted this template to describe two ways of approaching your practice of sense foraging—doing it either tired or wired.

TIRED: You're going bonkers staying indoors with your small child, so you drag your toddler to the park because "it's good for her." She seems happy in the sandbox, so you breathe a sigh of relief and collapse onto the park bench. You spend your time on your cell phone, catching up on email, but find that you can't really concentrate on the work. Your child keeps begging you to play with her, but you're not really there with her, either. But then she starts screaming about sand in her eyes, and your stress level skyrockets. You rush over to help, telling yourself, "I'm a rotten parent. But I only looked at my phone for a second." You arrive home feeling worse than when you left.

WIRED: You're going bonkers stuck indoors with your small child. You realize that you're building up resentment and feel stuck, so you pack up your toddler and take her to the park. She wants to play in the sandbox, so you get down with her and start tracing patterns in the sand. You notice the birds singing, the sway of the trees, and your child's glee as she lets the sand sift through her fingers and buries her feet. The wind blows and you both get a bit of sand in your eyes, but you quickly reassure her and wash away the grit in the nearby water fountain. You meet some other parents and have a nice chat, while your child makes a new friend. You arrive home completely refreshed.

Suggestions:

"Touching grass" requires at least two hours a week to produce a noticeable change in well-being. But when you consider that there are 168 hours in a week, that's not really such a huge investment. And there's no shame in going for the simple, mundane, and somewhat obvious approaches:

- Eat your lunch outdoors, near a fountain or a green space.
- Walk in the park or the woods. Inhale deeply. Notice the height of the trees, their movement, the way the sunlight filters through.
- Take off your shoes and walk barefoot. Reconnect with what the soles of your feet feel like.
- Set your alarm clock to catch a sunrise or make time to watch the sunset.
- Get an indoor plant; take some time each day to say hi to it, check its soil, water when necessary.
- Install a bird feeder and watch the birds gather. See if you can notice familiar visitors from day to day.
- Clean up trash in your neighborhood; there's often an unexpected sense of empowerment and delight in making a small difference.
- You might not have access to a pristine glade at work. Not every employer provides unicorns and panoramic vistas. But getting outside still matters. Get out of your office to go buy coffee, or come up with some other reason to take a walk every day. Even in a sprawling metropolis, there are opportunities to feel a bit of weather on your face or move through a different environment.

- Get out on a lake and go fishing; it is a fantastic opportunity to think about how your sensory environment predicts the behavior of wildlife around you.

We're fully aware that these suggestions are not rocket science. But one of the reasons we've presented the science behind sense foraging in such detail is to demonstrate that the simple actions you might dismiss with "Yeah, yeah. Take a walk. Smell the roses. Sounds like something my grandmother would tell me" actually have dramatic effects when taken to heart. It's not rocket science, but it *is* neuroscience. And it's been proven to work, time and time again.

FEEL THE BURN (EXERCISE)

Just as a sedentary lifestyle is a widely acknowledged risk factor for many health conditions, physical activity has been shown to reliably reduce that risk. A large population study of more than 360,000 people in the United Kingdom indicated that people who reported being sedentary between two and six hours per day had a 27 percent increased risk for a number of noncommunicable diseases, including heart disease, liver disease, thyroid disorder, depression, migraine, and arthritis.[7] The investigators then examined what would happen if that time were theoretically replaced with light, moderate, or vigorous physical activity. They found that the more vigorous the exercise, the greater the protection from the noncommunicable diseases described.[8]

At a neural level, exercise seems to make the brain more efficient. Brain activity increases during cognitive tests following exercise, both for attending to stimuli and for inhibiting inappropriate responses.[9] Efficiency alone is no guarantee of feeling better—one could become a more efficient ruminator, for instance. Yet mix in the ability to inhibit knee-jerk reactions and you have an important element in disrupting the bias toward active rather than perceptual inference.

If exercise helps people feel that they have more wherewithal in the face of stress, it can also give them more options for how to respond, in terms of both internal judgments and overt actions. With practice, increased responsiveness to one's environment and physiology with a sense of effortless control is an example of sense foraging often referred to as a "flow state." From our perspective, flow is putting in the time to develop an accurate model and then savoring as the model now takes care of you.

Sports journalism is rife with references to flow: an unfolding of skillful physical performance that does not seem to require effort or direction. Sounds a lot like presence: being in the moment, like Sully landing his jetliner on the river, with no extraneous thoughts. Baseball players refer to it as "see the ball, hit the ball." In basketball, it's called "lights out" shooting, or being "in the zone." Across these cases, we can see the mechanics of sense foraging at play—flow states use quick and accurate perception (perceptual inference) to dynamically match one's situation to a preexisting model. In true flow, no model updating is required—one's inclinations to respond are met without any resistance—there is trust in one's wealth of sport-specific knowledge, trust that if one sees a situation clearly, one can rely on responses hammered into the DMN by countless hours of practice and preparation. Flow is the sense of accomplishment that comes from having a DMN that is programmed to respond ideally to your situation. Perception leads to action seamlessly to achieve a desired goal.

Part of what makes flow so attractive during nominally effortful and tiring activity is that it introduces a sense of ease and pleasure that is atypical of extreme exertion. Many people have tried to explain what the flow state is and how it works. One idea is that intense effort causes endorphins, natural analgesics that reduce pain and provide a warm and cozy feeling, to be released within the body. But such explanations are not particularly satisfactory, because the magical thing about the flow state is that performance is held at an optimal level, whereas typically painkillers dull the senses and impair performance.

But here's a neglected insight: scientists have a lesser-known recipe for the flow state that sounds a lot like perceptual inference.[10] Research shows that the experience can be explained by two major factors. The first is a sense of positive engagement—a willingness and curiosity to throw oneself into the activity. But, counterintuitively, the second factor is not perseverance, effort, or how much intensity you can muster. Instead, the key component is tranquility, the feeling that one does not need to do anything more, that one can be engaged without having to strive or push beyond the behavioral repertoires that are already unfolding. The tranquility of the flow state comes from having your expectations so finely tuned that you're engaged and paying close attention to your interactions with the world, but none of what happens comes as a surprise.

Flow is associated with elite athletes and performers because their intense preparation makes it safe to let go of conceptual thought and immerse oneself in sense foraging. This allows the Olympic downhill skier to automatically take an accurate model of speed and terrain and get down the mountain safely. The rest of us might want more time to figure out how to respond! But the mechanics of sense foraging suggest that any situation can become a flow state, if one has sufficient preparation and the right mindset to get there.

You don't have to exercise this way; you can look at exercise as just a form of active inference—"something is wrong with my body and I'm going to fix it"—and you can simply go through the motions and it will still be good for you. But exercise that prioritizes flow can be a powerful avenue for sense foraging your way into agency and empowerment. For example, you might notice with experience that your mood isn't destiny—if you start exercising in a bad mood and notice how challenging your body introduces change to your physiology, you can form a new model (perceptual inference) for emotion regulation. Eventually, you can check in with your senses, notice a low mood, and feel confident that you can change it. If you achieve a flow state, you can even watch your mood improve as your body goes through its physical exertions.

TIRED: Imagine that you've begun a jogging routine. You start each run in a bad place. You're doing this only because you're fed up with being out of shape, and you dread the exertion, the boredom, and the possibility of sore knees. You force yourself to take the first few steps, which confirm your negative expectations, but you've felt this way before with exercise, and you know you'll push through. You find a way to get through your run mostly on autopilot, and you occupy your mind trying to figure out how to deal with a problematic coworker. You survive the run, realize that at least you feel a little more alert, and it's one day down. You shower and resign yourself to continue the routine.

WIRED: You still start off from a place of dread. But you remember to tune in to what is happening in your body. Halfway through, you notice that you're more focused, engaged in a moment-to-moment negotiation between your exertion level and feedback from your body. It becomes a kind of conversation, while big-picture thoughts and doubts about what you can or can't do become irrelevant. What matters is the next stride, the next hill, whether you should change your breathing to address the stitch in your side. You decide to push harder and begin to feel the endorphin release that creates the "runner's high." With all that oxygenated blood pulsing through your veins, you notice a broader sense of freedom and a changing relationship to pain and your ability to manage it. This physical accomplishment gives you renewed confidence in your ability to manage problems in general. As you shower, you notice that a lot of other issues in your life just don't seem as important, because you just experienced real evidence of your ability to stay engaged and work your way through a challenge.

Remember, though, that you reached this point not through grim determination and masochistic drive (special features of the House of Habit) but by staying in the moment to engage with the wide range of sense information streaming in. Simply by getting out of the way and surrendering your need to control the experience, you become a collaborator with your body rather than a slave to your DMN. By doing that, you can transform not just your physical fitness but your life in general.

Suggestions:

If you already have forms of exercise that you enjoy, those are probably good places to start the transition from physical activity to sense foraging. Here are some examples:

Light Physical Activity

Casual walking, stretching, dancing slowly, gardening, fishing.

And here's an offbeat idea. There are many chores around the house that must get done. Once you find yourself doing one, why not treat it as an opportunity for sense foraging and try to notice something new about the experience?

Moderate Physical Activity

Brisk walking, hiking, walking uphill, bicycling, low-impact aerobics, yoga, golf, more challenging housework that involves movement, like mopping, vacuuming, cleaning, scrubbing, shoveling snow, or lawn mowing.

Vigorous Physical Activity

Running, wheeling a wheelchair, backpacking, mountain climbing, elliptical machine, stationary bike, karate, judo, tae kwon do, jumping rope, weight training, basketball, soccer, hockey, swimming, downhill or cross-country skiing, occupations that require heavy lifting.

VINCENT VAN FLOW (ART)

The film *White Men Can't Jump* portrays an unlikely friendship between a Black man (Sidney Dean, played by Wesley Snipes) and a reckless, happy-go-lucky white man (Billy Hoyle, played by Woody Harrelson). The plot revolves around their shared love of basketball and their shared need to make some money. In one iconic scene, Billy tunes the radio to Jimi Hendrix, arguably one of the best rock guitarists of all time. Sidney is incensed, asking Billy what he thinks he's doing. Billy defends his choice, saying that he likes to "listen" to Hendrix. Billy's attitude is that music can just provide a quick hedonic hit—it is familiar to him, and he enjoys it. Sidney responds that Billy can "listen" but can't "hear" Hendrix. For Sidney, Hendrix's music is more than just light entertainment; he finds meaning and cultural connection with this iconoclastic Black performer.

Within a not-particularly-high-brow movie, a powerful philosophical claim is being made. A work of art, no matter how venerated it may be, does not change the viewer or listener all by itself. What matters is how sensory engagement is leveraged when encountering art—some people may prefer to stay at a surface level of enjoyment, whereas for others, appreciation means self-reflection and potential growth.

For each of us, works of art exist that go beyond a purely aesthetic experience to provoke unexpected and challenging thoughts and feelings. Art can be a powerful disruptor of the status quo, resisting the quick template matching of the DMN, making salient other aspects of our sensory experience. To let an external pattern affect us internally, and to be curious about it, is a wonderful recipe for sense foraging and subsequent perceptual inference.

Art that truly speaks to us can be a portal to sense foraging, allowing us to enter a world of novelty and surprise. Even the oft-panned single-black-square-on-a-white-canvas style of minimalist art can do this, not because of technical mastery, but because it prompts us to reflect on where value comes from and, perhaps in doing so, to question the arbitrariness of what we ourselves value. And even if we don't get philosophical about it, any work of art is essentially an exercise in sensory transmission. The question is, Will we just look or listen, or will we actually try to see and hear?

Art therapy has gained stature as a clinical intervention. For younger people or those for whom communication is difficult, it's often preferred to more traditional therapies. Research suggests that stimuli that are rated as beautiful evoke less frontal brain activity, which may indicate less cognitive processing.[11] Such findings are akin to "awe" experiences found in nature-based sense-foraging practice. Beauty can suspend judgment and promote sensory immersion.

The subjectivity of aesthetic experience does raise the question of what the minimal conditions are for art to trigger sense foraging. There's a big difference between a classical masterpiece and a pop song, between a film by Pedro Almodóvar and the latest offering from the Marvel Universe. Today, so much of media appeals to the appetites rather than the aesthetic sense, making it simply a way of "amusing ourselves to death." It's also designed to be habit-forming, which means that the DMN pushes us online as a reflex action. Some people use social media on their phones while streaming a movie—distraction from a distraction!

Two questions are critical for using film, music, dance, painting, or sculpture as a passage into sense foraging. Does the work provoke a reaction inside you? And are you willing to explore where it leads? Art works when it provides an opportunity to engage with uncertainty rather than simply enduring until the very predictable climactic shootout or rom-com happy ending.

TIRED: You go to an exhibition of a famous abstract artist's work and find yourself surrounded by people gushing. You stay silent because you can't see what all the fuss is about. You reflect (silently)

that your five-year-old could have painted these canvases, and you (silently) dismiss the fervor as another example of pretentious groupthink. You spend a couple of hours wandering the halls and find a few pieces you enjoy. You leave satisfied that at least you made the effort to absorb a bit of culture.

WIRED: You go to the same art show and get into a conversation with some of the people gushing. You ask them with genuine curiosity what they find so great about the work, and you're honest enough to express your lack of enthusiasm. You open yourself up to feedback and end up questioning why one person sees beauty while another sees a marketing gimmick. You agree that art is open to many interpretations, and while this work doesn't do it for you, you see the value in exploring your aesthetic beliefs, making room for a diversity of preferences. As you leave the gallery, you end up spending a full ten minutes absolutely wowed by a work of representational art displayed at the museum's entrance. You are floored by how a simple painting stirs up powerful feelings within you, and you take some time to explore this cascade of emotion.

Suggestions:

You don't have to travel to a museum to explore fine art. Most museum websites or even Google Images can give you access to high-resolution images of famous works. Search for something colorful that appeals to you, then challenge yourself to look more deeply into it. You could start with Monet's *Water Lilies*, for instance, or a Zulu beadwork bracelet. Set it up on your screen so that you can spend a few minutes concentrating on the piece. At first, you might notice familiar categories popping into your head—"There's a lot of blue in the Monet painting" or "The beadwork is very colorful." Now look more closely as you start to color forage. Can you pick out four different shades of blue in *Water Lilies*? How about four different shades of red? If you're looking at the bracelet, can you pick out the black beads that separate and contour each color? Can you pick out three geometric shapes? Color foraging allows us to appreciate that while these features were there all along, we might be seeing them for the first time.

If visual art isn't your thing, no worries. You probably have a sense of some media that speaks to you. Perhaps you can survey music, film, dance, or other performances that make you feel more deeply. The trick now is to reencounter one of these works with sense foraging in mind, almost as if you're seeing it for the first time. Can you take a deep dive into something familiar, only this time, make it the sole focus of your attention? Can you explore not just its external sensory features but also the feelings it provokes inside you, and how they unfold over time? There's no right answer. As you go deeper into sense foraging, you create the conditions for perceptual inference, and you set yourself up for change. Here are a few suggested activities to get you started:

- Play one of your favorite songs and savor something new in it. Maybe focus on the bass line or the drums.
- Alternatively, play a song from a genre you despise. Listen to it and explore all the reactions you have. What exactly rubs you the wrong way? Do you even have a reason beyond knowing you don't like it? Conversely, is there some redeeming feature that gets masked by a reflexive habit of rejection?
- Rewatch your favorite show or movie, and home in on what moments make this experience special for you.

Notice anticipation and fulfillment of your expectations for each scene.

- Create art yourself. Buy some colored pencils and start by drawing circles, triangles, or squares, filling a sheet of white paper, or color in the designs in a coloring book, or do this with watercolor or acrylic paints using heavier paper. Can you notice your color choices? What thoughts, feelings, or sensations inform your process of coloring or creating?
- Think of a time when you showed resilience in coping with a difficult situation. Now write a song or a poem about that. You can tell yourself that no one will ever see or hear it, and promise to burn the page afterward, but take a risk and express yourself. What did you go through? How did it feel? How did it end? Sometimes having to put things in a lyrical or poetic form can be more liberating than writing out the familiar internal monologue.
- Magic isn't fine art, but good magic is a skilled performance, and it provides an opportunity to sense forage. The experience of surprise, awe, and wonder represents prediction errors at work. Explore incredulity and feeling stumped as you watch a magician do something you believe is not physically possible.
- Watch TikTok dance reels that use one of your favorite songs as their soundtrack. See whether you can imitate the first thirty seconds of the dance routine.
- If all else fails, when was the last time you grabbed that hairbrush/shampoo-bottle microphone and sang in the shower, reveling in the acoustics?

HIT THE ROAD, JACK (TRAVEL)

Traveling for pleasure is another familiar access point for awakening the senses. Studies show that people experience a greater sense of meaning and a reduction of stress, even at a physiological level, after as few as three days of being somewhere new.[12]

The reason is obvious. We're talking sensory awareness, after all, and in a freely chosen new environment, every sight and sound and smell is different from what we're used to. We're also separated from the environmental cues that trigger habitual behavior. The desk where despair has set in by 3:00 p.m. every day? Far, far away. The alarm clock's buzz that pulls you from sleep into a state of resentment and weariness? Back on your nightstand at home! All those people who say unsettling things to you? Nowhere to be found.

When we're in familiar environments, the sensory details—even the pleasant ones—fade into the background, leaving us to focus on information that's emotionally provocative (generally speaking, threats and deadlines), while the default mode network supplies predictions about how to respond. In an unfamiliar place, we're far more likely to notice and appreciate the sunlight, or a breeze, or the chirping of birds, or the taste of food, or a good night's sleep.

But to be "transported" by travel, you must be committed to sense foraging, perceptual inference, and a new model for living. Otherwise, there's no guarantee that strolling a quiet street in Paris or hiking in the Rockies won't provoke the same DAMN narratives anytime you hit a snag.

Conversely, researchers have started to investigate whether less intense forms of travel, even the recent idea of the staycation, can create the same feelings of relief and restoration.[13] They found several factors that predicted a transition from distress to renewed hope, optimism, and resilience following half-day sightseeing bus tours of the local area. While participants generally felt better after such tours, benefits were much stronger for participants who had a strong motivation to escape from their routine, were open to novelty, and had a sincere interest in learning about their hometown. But these effects were all amplified for those who were able to see this admittedly local travel as an opportunity to take in new information rather than just to rest and escape. So, you don't have to break the bank to benefit from travel if your motivations are in the right place. All that's required is a desire to get unstuck and excitement about engaging the senses. It's not about just looking at the world—it's about really seeing it.

TIRED: You feel so stressed at work, you decide to get away. You drain your savings to get to some exotic place, only to find yourself annoyed by how gosh-darn *foreign* everything is. It's crowded, the food is strange (not to mention that the menu is in a foreign language), and don't even think about the toilet situation. You're on edge from feeling that you must constantly ward off the dangers and discomforts of the new environment. You wind up at McDonald's and later tell your friends that finding it was the highlight of your trip.

WIRED: You feel so stressed at work, you decide to get away. You hear about ghost tours in your own town and sign up for one, walking through unfamiliar side streets on your way to the meeting place. You discover details about the history of the neighborhood you never imagined, and you meet interesting people on the tour, some of whom are visiting from exotic locales. You learn all about the haunting of the big house on the corner, and from that day forward you get a shiver every time you pass by it.

Suggestions:

- Seeing the world is great, but if you're not an experienced traveler, we recommend you start off by sense foraging closer to home.
- Go somewhere local you've never been, specifically *because* you don't know what to expect.
- Take a local tour that focuses on something you aren't familiar with, like the historical ghost tour we mentioned above, or tours of local food, art, architecture, or wildlife. Most parts of the world have a tourism industry; we just tend to ignore our local version.
- Visit a nearby conservation area or historic site, particularly if you can get a guide to point out all the hidden details you might miss otherwise.
- Consider making a journal, or at least noting the things that surprise you on your trip, whether it is a three-hour local tour or a monthlong trek through a foreign land.

BE HERE NOW (MEDITATION)

Spiritual and contemplative practices such as meditation and yoga originated in Asia but have blossomed in the West, driven by people's curiosity and desire to settle their minds and bodies amid the ceaseless demands of modern life.

At their core, these practices explore the same concept—directed attention. The starting point in meditation and yoga involves being able to notice where our attention is, to recognize that we can bring it back to the present moment, and then to explore what we notice without judging what we find.

By stepping away from frames of reference ingrained by the DMN—evaluating, comparing, avoiding, trying to fix, trying to achieve—we're able to see other aspects of our experience that were previously obscured. This is why meditation is considered the original approach to perceptual inference.

While the term "meditation" refers to a variety of practices, including tai chi, qigong, and yoga, all progress through a similar principle of mental training to expand awareness beyond a conceptual state of mind to reveal one's experience more directly. All these practices encourage sense foraging in the moment, suspending goal orientation in order to explore the target of attention, whether it's the feeling of breath and body in mindfulness, movement through a series of yogic postures, or a sense of rootedness and connection to the earth in qigong. These practices also provide some of the benefits of exercise we discussed earlier. But the great thing about meditation is that it provides access to sense foraging even when physical movement isn't possible.

Let's take a simple practice of meditation where the general guidance is to "pay attention to your breathing." Because breathing is under automatic control, we generally pay very little attention to it, except perhaps if we run into respiratory difficulties. It's more common to *think about* our breath than it is to feel ourselves breathing in each moment. But breath meditation invites us to explore each inhalation and exhalation, attending to sensations caused by the air entering the body and leaving the body. No longer tied to the idea of "the breath," we become aware of the rising of the chest, the lifting of the abdomen, the friction at the nostrils, each a part of a moving, living thing. We also become aware of how difficult it can be to pay such close attention for more than a few seconds.

Research has borne out that one big thing mindfulness and other contemplative practices do for us is to build patience. By this we don't mean patience with a capital *P*, one of the seven virtues opposing the seven deadly sins. In behavioral studies, "patience" is common parlance for how well someone can inhibit a response deemed to be inappropriate. The go/no-go task, designed to measure this version of patience, asks participants to press a button when they see one stimulus (e.g., "O") on a computer screen, and not to press when they see another (e.g., "X"). Fewer errors mean better response inhibition.

Are meditators better than the average person at response inhibition? Neuroscientist Catherine Andreu and colleagues used athletes as a control group to test this question on an emotional go/no-go task using positive, neutral, and negative pictures.[14] Athletes are supposedly very good at this sort of thing, because the mental aspect of sports training leads to various improvements in cognitive function, including inhibitory control. Yet meditators made fewer errors than the athletes and were better able to inhibit responses when the no-go signal appeared.

A late-fourteenth-century definition of patience as "quiet or calmness in waiting for something to happen" fits very nicely with these findings.[15] The practice of meditation provides this ability to pause, exert some control over preprogrammed reactivity, and choose what to do next. This kind of patience allows us to insert ourselves into the next moment, creating room for our values and intentions, rather than charging ahead with default behavior dictated by the DMN.

TIRED: You've heard a lot about a highly rated meditation app and download it on your phone. You force yourself to practice, even though it's uncomfortable and you often wonder what you're getting out of it. Nevertheless, you grind out thirty days of guided audio meditations and hope that it "fixed you."

WIRED: You explore a variety of contemplative practices, including formal, seated meditation, as well as movement-based practices like yoga and tai chi. You find the style of practice that speaks to you and hooks you into wanting to do more. You're not sure it's "fixing you," but you feel less stuck. You also feel that you're beginning a journey, and you become increasingly curious about where this journey will take you.

Suggestions:

If you're going to go the meditation / contemplative practices route, you're going to need guidance. On the surface, these practices may seem deceptively simple—focus on breathing in and breathing out, walk slowly and feel every footfall, practice stretching your body with others in silence. Yet, guidance is what will unlock their deeper transformative potential for you.

Our backgrounds are primarily in the mindfulness meditation tradition, so we'll limit our examples to that kind of practice, knowing that the basic precepts can be applied to other approaches as well.

Solo, single-session exploration: YouTube can be a portal for you to find teachers who will take you through a yoga, tai chi, or qigong class; other teachers will lead you through practices for mindful eating, walking, loving-kindness, or sitting with the breath. All you have to do is show up and press play. One caveat is that what's freely available online can vary in terms of quality and expertise. Maintain a healthy skepticism while you sample different sites.

Solo, multisession exploration: Meditation apps or books can add structure and expert guidance. While we're not in the business of endorsing specific apps, we do favor those that feature experienced meditation teachers. These include Insight Timer, Ten Percent Happier, Calm, Headspace, Jon Kabat-Zinn Meditations, and AmDTx. There are also delightful books to help you understand both theory and practice, such as *Wherever You Go, There You Are*, by Jon Kabat-Zinn, *Lovingkindness: The Revolutionary Art of Happiness*, by Sharon Salzberg, *The Path of Insight Meditation*, by Joseph Goldstein and Jack Kornfield, and *Peace Is Every Step*, by Thich Nhat Hanh. There are also meditation guides for more specific issues such as mental health or parenting, like Segal's own collaboration *The Mindful Way through Depression* (with Mark Williams, John Teasdale, and Jon Kabat-Zinn). We recommend you start with the general and then tunnel into specific topics.

Learning with a teacher: If you're getting something from an app or a book and want to go deeper, live classes allow you to learn with greater support and personal feedback. Structured programs such as mindfulness-based stress reduction (MBSR) and mindfulness-based cognitive therapy (MBCT) are eight-week group programs in which you learn how to practice mindfulness and build greater resilience and well-being. The Mindful Directory (mindfuldirectory.org) is a no-cost approach for finding MBSR, MBCT, and other mindfulness courses in your neck of the woods. Alternatively, if you wanted to access one of these programs online, you could check out eM Life (emindful.com), the Oxford Mindfulness Foundation (oxfordmindfulness.org), the Centre for Mindfulness Studies (mindfulnessstudies.com), or the Mindfulness

Center at Brown University (brown.edu/public-health/mindful ness). To look even further afield, the International Mindfulness Teachers Association (imta.org) lists teachers whose teaching is not tied to these eight-week structured programs.

Maybe the time-honored tradition of working with a teacher at a meditation, yoga, or tai chi center appeals to you. It may be daunting to think of going to a temple, meditation center, or community center to explore something new, but most of these centers also offer introductory classes or programs that are designed to provide a gentle introduction and let you decide for yourself. Take a look at some of them and see what you think: Spirit Rock (spir itrock.org), Insight Meditation Society (dharma.org), and Gaia House (gaiahouse.co.uk). If you would prefer to have the live experience online, there are also great virtual resources—for example, You Aligned (youaligned.com), or the Upaya Zen Center (https://www.upaya.org/self-paced-programs/). And increasingly, local teachers might offer online sessions if the commute is too much. Again, we're not endorsing any particular site, but rather helping you to begin your own exploration.

FAR OUT, MAN! (PSYCHEDELICS)

Nearly thirty years after the state of California banned psychedelics, and the world followed suit, research on these mind-bending substances resurfaced in the mid-2000s, buoyed by surprising scientific findings. Scientists studying the drug ketamine, a horse tranquilizer turned dissociative psychedelic and party drug (Special K), made the startling discovery that ketamine promoted neurogenesis—brain-cell growth—in rats. And all this time we thought that "our brain on drugs" looked like an egg in a frying pan. Adding to the surprise, scientists at the time doubted that new cell growth was even possible in adult brains. Ketamine dispelled both misconceptions, showing clear evidence that your brain can grow new neurons if given the right push.

Most recently, MDMA, the pure form of the rave drug Ecstasy, or Molly, was approved for treating PTSD in veterans. Encouraging results led to partnerships between prominent medical schools like Johns Hopkins and the National Institutes of Health, and similar relationships were established in the United Kingdom at Imperial College London. The results were staggering. Patients with treatment-resistant PTSD, anxiety, and depression began "waking back up" to their lives. These were folks authorized to try ketamine, MDMA, magic mushrooms, or LSD because they had run out of other options and seemed doomed to live with chronic impairment. And suddenly the medical news was full of claims that one could have a "top 5 lifetime meaningful experience," guaranteed, in one supervised trip.

While research in this area is still in its infancy, there are powerful forces promoting acceptance of these drugs, with huge amounts of money being invested. We believe that sooner or later psychedelics will be legally prescribed to help with your mood or concentration. So . . . should you take the plunge? Can psychedelics be your point of access for sense foraging?

In the right context, yes. Psychedelics have the potential to increase communication among different parts of your brain,[16] which translates into more intense sensory experiences of all shapes and sizes. Ketamine, that old horse tranquilizer, was already associated with improved mood and reduced suicidal thinking; we now know that parts of these benefits come from helping people to update their beliefs, increasing their ability to integrate positive interpretations into what were previously hopeless models of the world.[17] Ketamine is already becoming legal to prescribe in many parts of the world; if a person with very rigid depressive models can get unstuck using ketamine, could it also support sense foraging more generally?

While these drugs do a lot of the work in helping to erode the hold of the DMN over our attentional focus, the basic principles of sense foraging still apply. To get the most out of a psychedelic trip, you might commit to using your trip as an opportunity for growth, with the clear goal of immersing yourself in sensation. You might do what those benefiting from ketamine treatment do and consider integrating different aspects of your experience into your internal narrative, finding moments of positivity amid well-rehearsed knowledge of stressors and challenges.

It is important to have a foraging intention when using psychedelics, because increased brain communication can cut both ways: the same lack of control that lets you engage in intense, unadulterated sensation also robs you of control if what you experience starts to spiral into darker corners of association. For this reason, it is strongly recommended that you have an experienced guide with you on a psychedelic trip. A guide can act as a sounding board for your experiences, to normalize what is happening and help prevent you from becoming completely untethered. Moreover, a guide can help rein in the DMN's tendency to influence your perceptual reality. If you end up merely reenacting your own worries, fears, or traumas, without any support to embrace sensation over interpretation, you have the recipe for a bad trip. At minimum, resources like www.tripsafe.org provide valuable psychedelic education to reduce potential harms.

The other major issue with psychedelics is that they are still largely illegal. In some parts of the world, ketamine is an off-label drug, which means that it's legal for some purposes, and a doctor can prescribe it for other uses. There are a growing number of clinics and research trials on psychedelics, and this is the avenue we recommend. Otherwise, psychedelic use would mean breaking the law, which means there's little oversight and quality control. Just because you order magic mushroom capsules by mail doesn't mean you're getting what you paid for. In our own research, illegality and lack of quality control end up being two of the chief stressors for people who try psychedelics.[18]

Still, in the right environment, with a skilled guide and some assurance of both drug safety and your own safety during your trip, psychedelics can do a lot of the sense-foraging work for you, like throwing you into the depths of the forest from the comfort of your couch. For those struggling to find the motivation and concentration needed to develop sense-foraging habits, and especially for those who feel deeply stuck in one way of perceiving reality, experimenting with these drugs can be like a splash of cold water on a sleeping face. Just seeing how radically different you can feel, how differently you can relate to the world, can help unstick you from a highly automated and rigid way of knowing.

TIRED: The wrong way to approach psychedelics is to try a bunch of drugs, often accompanied by alcohol and marijuana, leading to blurry and confused experiences, losing track of your values, possibly getting into trouble wandering about while hallucinating, and no clear benefit.

WIRED: Approaching psychedelic use intentionally can give you a chance to explore the different ways your senses can unfold when they are less constrained by learning. To do this, you need to find a safe space and an adequate amount of time. If you are not experienced with the drugs, you need to find someone experienced to help guide you, with whom you can predetermine the type of support you'd like. You also need a clear, anchoring intention to come back to sense foraging rather than completely breaking from the world into hallucination and delusion.

Suggestions:

We do not recommend that you break the law. There are, however, some legal options:

- Get to know what drugs are legal in your area. In Toronto and New York City, for example, you can get prescribed off-label ketamine.
- If you find a legal drug in your area that requires a prescription, look for local clinics that offer one-stop assessment, referral, and trip guidance.
- There are also legal psychedelic retreats in many parts of the world where drugs are not criminalized. If you're willing to do some research and then travel, there are highly rated clinics or retreat centers in the Caribbean, Mexico, and the Netherlands.
- We cannot in good conscience suggest that you perform any illegal activity. Such actions are not trivial, and in one large study of psychedelic users, feelings of anxiety or guilt over breaking the law (and the repercussions of getting caught) are among the most frequent downsides to psychedelic use. If you are pursuing psychedelics to deal with anxiety or worry, doing so illegally may substantially reduce net benefits.[19]
- For experienced psychonauts, feel free to set your own rules, but be aware of the importance of setting.[20] Avoid

being in situations in which you'll be influenced by other people who may not be aligned with your foraging intentions.

WHAT'S YOUR ACCESS STYLE?

We've provided a survey of potential access points. What's important is not finding the "right" one but finding the one that's the best fit for you. Part of that fit is selecting the right activity, but part of it is also engaging in that activity in a way that supports your learning style.

We all know the extrovert/introvert personality distinction, for example, and there is a ton of data to support it. Some people simply like being around and learning from others, while some find social situations draining and treasure their personal time. Some like to feel that someone is holding their hand and offering feedback on each exercise; others hate that kind of interaction but love puzzling things out on their own. As you consider your own access points, it might be worth asking how much a teacher or guide should be part of your process.

While sense foraging is a deeply personal practice, many of the access points mentioned here were first recognized for their value in highly supported contexts. Many contemplative practices were developed in monastic settings, with mentorship and mutual support, then refined into formats that could be scaled to engage the broader population. Similarly, most psychedelic substances were originally used as sacred medicines, part of rich cultural and religious traditions to support community wholeness and a feeling of connection to the natural world. And indeed, critics of the commercialization of these access points worry about the solitary and secular pursuit of them as having missed the point. The reality is that there are pros and cons to having other people guide you in sense foraging. It does make sense that for some aspiring foragers, support from someone you consider an expert will be helpful. For others, support may be less important if the proper context and mental preparation are in place.

The other major consideration is ease of access. These days people tend to assume that value and price are synonymous—if it's cheap, it can't be good, and if it is expensive, it must be worth it. Don't feel that you must do something drastic and expensive to start sense foraging. The principles of *ubiquity* and *immersion* are all about the idea that foraging is almost always available to you. The following table illustrates the variety of access points organized by both levels of support and accessibility.

	AUTONOMOUS		**SUPPORTED**	
	High Access	**Low Access**	**High Access**	**Low Access**
Nature	Savor a walk in the park	Climb a mountain	Tour a botanical garden	Forage for mushrooms with a guide
Exercise	Take a brisk walk in the park	Surf	Join a neighborhood running group	Circuit train with a coach
Art	Color in adult coloring books	Visit the Louvre	Sign up for a pottery or wine-tasting class	Take an in-person masterclass
Travel	Visit a tourist destination in your own town	Visit a tourist destination in a foreign land	Take a historic tour of your local neighborhood	Go on a guided trek to a remote part of the world
Meditation	Engage in concentrative practice (e.g., breath counting)	Take a personal retreat (a longer period of intensive, self-directed meditation)	Use a guided meditation app or recording for meditation, yoga, tai chi, and the like	Take a meditation course (MBSR, MBCT) or a silent retreat

Psychedelics	Take an unguided trip at home	Take an unguided trip at a retreat or foreign clinic	Take a guided trip, off-label (e.g., a ketamine clinic)	Take a guided trip at a retreat or foreign clinic
You choose! ________	*Something you can do yourself right now*	*Something you can do yourself with some planning*	*A class, guide, or group you can find right now*	*A class, guide, or group you need to find and schedule*

So, what's your style? Do you like working in groups or on your own? Do you feel like you already have a sense-foraging practice that you can dive into, or are you unsure how to get started? What's your level of access? Is nature right outside your door, or do you need to organize transportation and lodging? We suggest you prioritize high-access activities to get started, regardless of your support preferences. Travel is great, but if you must wait a few years to make a trip work, what are you going to do in the meantime?

Exercise: Commitment to Foraging

The exercise to finish this chapter is open-ended. Pick at least one access point from those explored previously that you can commit to exploring this week. Or, if you have an activity that helps you move into a sense-foraging mode that isn't listed but still adheres to the principles we've discussed, work with that. Pick one specific access point, and be *explicit* that you're using that activity to access the sensory world. The key is to transform an ordinary experience into an act of witnessing—riding a bike, I feel the growing fatigue in my muscles, a sense of uncertainty about finishing the interval, a change in breathing rate; even feedback from my heart-rate monitor is dynamic and tied to my bodily exertion in each moment.

So, here's your chance: *write down one sense-foraging practice that you will try in the next week*. Please make sure it is something safe and available to you, and that it provides an opportunity to immerse yourself in the *not knowing* of sense foraging. We encourage you to do something *new* if possible, rather than rehearse a practice with a lot of familiar expectations.

This week, I have chosen ______________________________ ____________ as my sense-foraging exercise. I will spend at least ______________________ minutes this week practicing my foraging skills. The biggest thing that could stop me from doing this is ______________________________, but I think I can manage to get it done because ____.

9

NOVICES SENSE, EXPERTS TOGGLE

> You have to stand for something or you will fall for anything.
>
> —ADAPTED FROM GORDON A. EADIE

SENSE FORAGING CAN HELP YOU ESCAPE FROM THE HOUSE OF Habit, providing a measure of relief by loosening the grip of rigid ideas and views of what is possible. While it is tempting to think of this as a self-help binary—the House of Habit is the problem and sense foraging is the solution—this type of black-and-white thinking fails to tell the whole story. Like it or not, the House of Habit is where meaning is found; renovation, and perhaps even fumigation, is needed, but in the end you must make peace with where you live—and for human beings, that means living through conceptual models of the world. *Toggling* between habit and sensation means learning to find balance between these two fundamental modes. Whether you are steeped in sense foraging to weaken rigid beliefs or returning from foraging with new ideas, it is critical to keep an eye on where you are headed.

Toggling reminds us that staying in one mode or the other is not going to allow us to predict the world more accurately. If sense foraging means saying yes to your sensory experience and weakening habit, toggling means making sure you are bringing back new meanings and forming the kind of habits that let you shout "yes!" to life as it unfolds. This may seem tricky, but we've been learning to strike this balance since our earliest years.

WHAT DOES TOGGLING LOOK LIKE?

It is amazing what little kids can't do. Despite being able to forage for all manner of sensory information when their caregiver is in a hurry ("Look, a ladybug!"), young children are horrible at finding lost objects. When a hat is not in its usual spot, parents can only shake their heads at their precious little darling's abject inability to recognize that same hat sitting on a shelf a few feet away. In similar fashion, children below the age of four are wonderful at exploring their sensory world, and are masterful at learning new rules, building basic models of the world. But they are horrible at reengaging sensation to update existing models when their hard-earned knowledge is revealed to be inaccurate. For example, a child who decides early on that they dislike a particular food might refuse to even consider eating other dishes that contain it, even when it is prepared differently; sometimes these models persist into adulthood, even though other tastes and preferences dramatically change.

Developmental psychologists in the 1990s explored this curious disconnect between children's excellent model-building skills and their surprising difficulties in updating these models. By the time a child is three years old, they are great at learning (and lecturing you) about simple rules, for instance, where the hat is supposed to go. But it isn't for another year or two that children can *update* their hard-earned knowledge and learn to shift between sets of rules. So, when the model says the hat should be in location A, it is very difficult for a three-year-old to update this model to entertain the possibility of looking in location B.

This inability to update models was elegantly illustrated by Professor Phil Zelazo and colleagues using the dimensional card-sorting task. They asked children to sort cards into two different piles, according to either shape (triangles vs. circles) or color (red vs. blue). Children easily learned to sort cards by red versus blue *or* to sort cards by triangle versus circle. Yet once the kids had learned to sort cards using only one dimension (e.g., color), three-year-olds struggled to shift to the other (e.g., shape). A child used to sorting by color can recite the new shape rule aloud but often fails to then change their sorting behavior accordingly. By contrast, most five-year-olds were able to switch between sorting rules on the fly, toggling into a model-updating mode and then exploiting a new set of rules to succeed in the task.[1]

Put simply, the ability to form and exploit models of the world precedes and then impedes our ability to update those models. And yet, the ability to toggle from exploiting existing knowledge to model updating is a universal, if later-developing, capacity. Indeed, the life of a child is characterized by continual calls for model updating as their world expands. Parents, educators, and natural consequences constantly demand this adaptation. At some point, the tooth fairy and Santa make way for more realistic accounts of where gifts come from, and we learn that hiding under the kitchen table won't keep parents from seeing us gobble up all our Halloween candy.

The ability to update continues to develop throughout childhood. Adolescents are better than children at tolerating situations where the outcomes are uncertain—they are objectively more comfortable foraging for novel solutions in unpredictable learning tasks.[2] Adolescents are also better at learning from their mistakes when the rules for success change, and what was previously rewarding ceases to be so.[3] Surprisingly, adults are more likely than adolescents to get stuck—the experimental equivalent of continuing to dine at a familiar restaurant even though the food's no longer so great.

Yet at the same time, adolescents are more reactive to stress—an affinity for sense foraging can be a double-edged sword.[4] Perhaps protecting ourselves from such reactivity is part of why we seek out certainty as adults, trading away our capacity for wonder and surprise to help us deal more effectively with the increasing responsibilities of adulthood. And so, we prioritize the sense of security afforded by the DMN's House of Habit. Yet this comfort comes at a cost: while we're probably better than a three-year-old at updating mental models, it is debatable whether we're better than a five-year-old at toggling between exploration and exploitation.

Imagine driving to a familiar location. Your regular route takes you off busy main streets and gets you to the office in less time. But one day, road construction turns your time-saving hack into a nightmare. You *should* be able to plot a new route and avoid the construction, and with some effort you succeed. Yet if you are distracted, you'll find yourself back on your old route and wondering why. You knew it was going to be clogged up. But ingrained models of the world have a life of their own that can be hard to extinguish.

As adults, even though we have the capacity to update our models, we often lack the motivation. In fact, we often resent being asked to change our minds about anything; when confronted with our own errors or contradictions, we sometimes want to shoot the messenger. Our partner tells us we're going to be late, and we shout back "I know! I know!" rather than acknowledging the need to rethink time management. We rail against another driver who accelerates through a yellow light and blocks our opportunity to complete a turn, but we rarely consider how many times we've done the same thing. Updating our models to acknowledge our own culpability is hard. It's so much easier to assume that someone else is to blame. We may have the ability to update our models, but most of us are not used to toggling into an exploratory foraging state and returning with a new set of rules.

In certain individuals, the inability to update models can be so extreme as to indicate brain damage. There is an adult analogue of the child-oriented color/shape sorting task known as the Wisconsin Card Sorting Task (WCST), and you can try it yourself at brainturk.com/cardsorting.

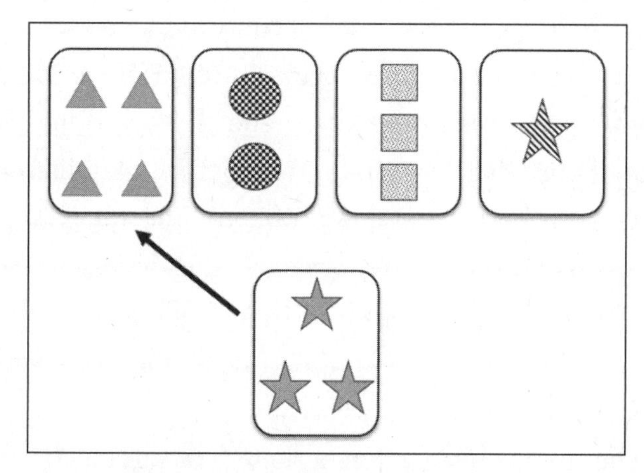

An example WCST Trial, sorting by "fill" rather than "shape" or "number"

In this task, you're asked to learn how to sort cards based on one of several properties (color, shape, or number of objects on the card). Once you figure out the rule—say, sorting is determined by color—it will work for a while, but then the rule changes. You then need to *toggle*, in this case letting go of the color category, and explore until you find the new rule (shape or number), then *toggle back* to start exploiting that knowledge again. How long do you stay with a rule once it's no longer working?

People who have damage in the front of their brains can often discuss rules in a rational way but lose the ability to update them. To be functional human beings we need to be able to make sense of the world, which means having mental models but updating them when appropriate.

TOGGLING AT THE HEART OF IT ALL

Part of our confidence in discussing toggling as the ideal mixture of perceptual and active inference is that philosophers throughout time have described these two capacities as intrinsic to human life. In *The Birth of Tragedy*, the German philosopher Friedrich Nietzsche describes two opposing forces of nature as the logical, ordered, and structured Apollonian, and the emotional, sensate, and frenzied Dionysian. Perhaps you recognize this distinction as a front-versus-back-of-the-brain problem—the tension between Apollo's reassuring, stabilizing force and the sweet chaos of Dionysian boundary dissolution. It is as though modern neuroscience has somehow mapped itself onto the same distinctions articulated in Nietzsche's reflections on antiquity.

Nietzsche saw the integration of these forces as an ideal, but one that was rarely achieved. Instead, he bemoaned how life's demands drew more on reflection and rule-based logic, which in turn crowded out lateral thinking and unconventionality. Nietzsche uses Greek tragedy as an example of successfully toggling between these systems. Characters in these ancient dramas often follow Dionysian arcs, ruled by their passions and desires—exemplars of sense foraging. But the Greek chorus, which dispassionately explains the character's machinations, helps the audience explore conceptual models for human behavior rather than being completely swept away by the drama. The magic of this theatrical convention lies in providing the audience with a framework for toggling between the Dionysian and the Apollonian. These performances, which originated as religious rites, provide a venue for sensing along with a character's specific and passionate experience, while also providing space to step back and explore a more general moral principle.

As with most philosophers, Nietzsche offered his thoughts and left it up to others to fill in the real-world implications. The problem with Greek tragedies is the moral lesson always follows great loss—that's why they call it a tragedy. Ideally, we should be able to make sense of our life before it descends into catastrophe. We need to discover our personal values, the things we're willing to work and fight for—the things that make life worth living. This means rejecting recipes for success others try to impose on us and finding our own, but this involves more than sense foraging to circumvent the DMN. We need to leverage sense foraging to discover our own recipes for living.

THE STEW THAT IS YOU

The tantalizing smell of beef and spices wafts from a row of crowded shops in central Bangkok. Its source is a huge cauldron of soup simmering inside the Wattana Panich restaurant. But this is no ordinary stew. Its unrivaled depth and complexity come from the fact that the pot's been simmering for more than forty years! Around the world, many cultures enjoy perpetual stews, adding ingredients over days, months, or even years and replacing only what's been eaten. Along with giving the food a deep, rich flavor, this process helps maintain a sense of continuity in the neighborhood.

Perpetual stews come from an age when starting the next fire wasn't easy, lean times were frequent, and there was no guarantee of fresh food each day. The stew helped to stretch the bounty of successful hunts and foraging expeditions and to create a consistent fare that drew from, but also supported, a whole community.

It turns out that this culinary curiosity is a good metaphor for how the DMN provides our sense of meaning. Scientists have begun to realize that even the most familiar and persistent truths about our self and the world are assembled on the fly each time we experience them. These "facts" are the product of whatever ingredients are on hand—much like that simmering stew in Bangkok. Unlike a computer, with fixed parts (processor, hard drives) but changeable contents (data), both the inputs (ingredients) and outputs (flavors) of the perpetual "self stew" can change. It appears there is no single recipe for making the feeling of you, and it's not written down in any cookbook. The feeling of who you are is no more and no less than the ingredients tossed into the cauldron as they blend together at a given point in time. The feeling of "you" is the stew served up by the DMN at any given moment.

The idea of our reality emerging in each moment emerged, itself, over a period of decades. William James, said to be the father of modern psychology, came up with the concept of stream of consciousness in 1890. If consciousness is a stream that sometimes contains feelings of "you" within it, then who we are is not fixed. Instead, the self is continually changing, with different elements (flavors) emerging at different times. As James wrote, "The I . . . cannot itself be . . . an unchanging metaphysical entity like the Soul, or a principle like the pure Ego, viewed as 'out of time.' It is a Thought, at each moment different from that of the last moment, but appropriative of the latter, together with all that the latter called its own."[5]

More than a century later, we can see that James was onto something. Our brains need to be active for us to have any sort of reportable experience, and the activity of the brain in each moment is an adaptation to the moments that came before. We take each experience, each recognition of success or failure, praise or judgment, and add it to the pot of self-relevant knowledge. It is only in this moment, as we reflect on the sum of all that has come before, that we gain a sense of self; a feeling of identity in this process of reflection. But the reflection must be ongoing to create a sense of "me."

Modern neuroscience supports this characterization. The DMN provides a sense of consistency to the types of things we think about ourselves and the world, to the point that we feel like there is a real "self" inside our heads and that we have come to know the true world beyond the boundaries of our bodies. But the sense of constancy is an illusion—the brain never forms some fixed "self." Instead, what we know about ourselves is simply our current model for predicting our own feelings and behavior, given to us effortlessly and automatically by the House of Habit. But there is no exhaustive list of properties that define us stored in our brains. Each moment of thinking "I like this" or "this isn't for me" is an act of comparing memories to the sensed environment and constructing a judgment of fit on the fly. The construction is not *arbitrary*, because it is based on our experience, but neither is it *immutable*—there is wiggle room in what memories we focus on and how much we prioritize consistency over change.

The fact that we are continually sampling the "self stew" for our sense of identity opens the door to personal transformation. Gradually, we can alter the stew's flavor by changing the ingredients. But what we can't do is just pour it out and start over. The stew is what's on the menu—there's nothing else. Just as meditators in our fMRI studies never really got rid of the DMN's self-referential processing but instead added to its repertoire, none of us is going to just swap out our brain's knowledge system. We must work from where we are.

LEARNING TO WORK WITH HABIT

The chrome of two motorcycles is glistening in the sun as the two riders mount their bikes. One downward flick of the wrist produces the throaty roar of their engines, and a backward heel stab lifts their kickstands off the ground. The two quintessentially American antiheroes are ready to hit the road. In a final gesture, one of them casually removes his watch from his wrist and throws it away. No longer tethered to time, he is free to live outside its demands for punctuality and responsibility. A new way of life awaits.

Two decades later, Thelma and Louise set out for a weekend fishing trip, leaving behind Thelma's overbearing and unfaithful husband, Darryl, and Louise's hectic job as a diner waitress. Before jumping into Louise's Ford Thunderbird convertible, they take a Polaroid of their new adventurous selves, preparing to leave their routine life behind.

And yet another two decades later, audiences caught a last glimpse of Don Draper, the advertising wizard from *Mad Men*, having given away everything he owns, sitting cross-legged on a hillside in California chanting "om."

These iconic scenes from the movies *Easy Rider* and *Thelma and Louise* and the more recent television show *Mad Men* clearly form a cliché that has stood the test of time. Each spoke to a generation's common understanding that for a character to grow and change, to complete their heroic arc, they must escape the strictures of habit into a wilderness of new sensation.

But lost in the excitement of sense foraging, there is a caveat. These three cinematic experiences showcase the escape but not the return. Most of us would prefer a round-trip ticket, and more conventional tales of self-discovery involve the hero going out to explore but then reintegrating.

There's a seductive element to these tear-it-all-down stories, but the happily-ever-after variety are more comforting, with the completion of the narrative arc, when the hero returns home, changed for the better. Dorothy wakes up at her Kansas farm after her journey through Oz, a more empathic version of herself. In *Lord of the Rings*, Frodo comes back to the Shire and feels empowered to address the corrupt denizens who have taken over. Proper heroes grow through their sense foraging, then find a way to use these experiences to improve their regular lives.

Completing the full arc of our own hero's journey requires learning to toggle into sense foraging and then back into the (slightly modified) world of habit. Toggling means leveraging sense foraging to update one's conceptual models but using it for personal growth rather than simply for escape. Recent neuroscience research helps clarify that sensation and habit don't need to have an antagonistic relationship. Our brains are wired to incrementally update habits by contributing new information to our well-established routines.

Consider a young couple embarking on the daunting process of buying their first house. They calculate mortgage rates, agent commissions, and the down payment to come up with a budget, then start ranking available houses based on the features they want: exposed brick, hardwood floors, a deck for entertaining, and lots of trendy restaurants within walking distance. But when they learn they're expecting a child, they revise their criteria toward greater square footage, multiple bathrooms, nearby playgrounds, and a good school district.

Michael Mack and colleagues examined toggling in the brain using a creepy-crawly version of the Wisconsin Card Sorting Task.[6] Participants were asked to categorize pictures of insects according to leg size, mouth shape, and antenna length. But then the rules changed, and participants had to learn a new set of sorting criteria. When participants had to update their models, the brain's dorsal attention network became connected to the hippocampus, a part of the DMN that supports our sense of environmental context. As participants correctly learned the new sorting rule, the DMN began to operate without the dorsal attention network's influence again, and the hippocampus now changed its response to the insects to reflect the new rules. The participants were now free to rely on automatic responses afforded by the DMN, with no need for effort or further updating from the prefrontal cortex.

This research gives us confidence in the idea that our habits can be updated with only a little bit of effort. While the DMN provides a model for what to expect from the world, it works less like a set of commandments carved in stone and more like the perpetual stew, something that can be updated to accommodate our own changing understanding of the world. The ease with which we can leverage sensation to update our mental models reminds us that sense foraging is not an endpoint. Life goes on; Redzepi gets back in the kitchen; Archimedes and Newton get back to inventing. Foraging provides a lot of new materials, but you must still assemble them for any kind of sustained value going forward. Jack Kornfield, a leading meditation teacher who specializes in engaging others in sense foraging, put it this way: "After the ecstasy, the laundry."

COMING BACK FROM THE MOUNTAINTOP

In brain terms, Kornfield's "ecstasy" likely refers to the brain quieting down enough to provide relief from the demands of always seeking to be in control. It fits nicely with the idea of escaping from the DMN's hyperactive House of Habit—the feeling of relief akin to putting down a heavy load. It is the "laundry" part that is harder to understand—why would a person go back to mundane household chores after tasting radical freedom? Our preceding discussion provides the answer. Most of us aren't trying to escape entirely from our jobs, relationships, and identities—we are just trying to live life in a way where we are less broken down by them. While a novice sense forager escapes from habit and thinks, "This is it, life will never be the same," the expert realizes that life goes on even after powerful moments of insight and model updating.

Yet what this looks like from a neuroscience perspective had not been understood. With researcher Cynthia Price, we designed a brain-imaging experiment to compare how novices and experts escape the House of Habit.[7] We hypothesized that experienced sense foragers would access deeper and more intense levels of sensory representation, getting those sense factories to work overtime. Processing visual stimuli is something we do every day. What happens when we break from routine and turn attention inward, toward the breath?

We outfitted participants with a keypad and asked them to do one of two things while in the fMRI scanner: either track their

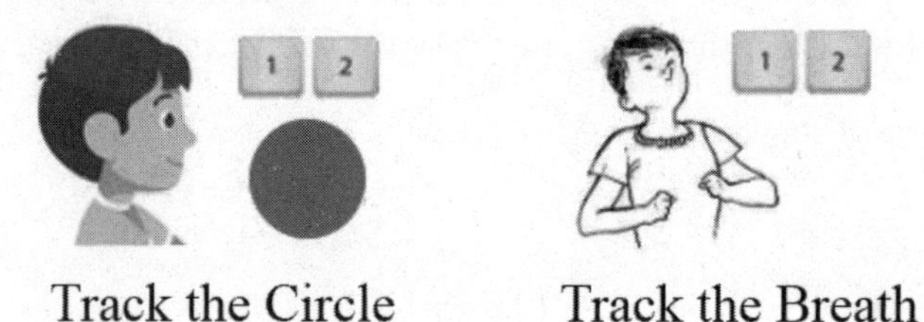

Example of the instruction screen for the experimental task

breathing or track a pulsing circle on a viewing screen.

As has become a pattern unto itself, the results shattered our own DMN-supported expectations. When participants focused on their breathing rather than the circle, much of the brain became less active. Participants were as accurate at tracking their breath as they were at tracking the circle, but it seemed as if they were doing so with much less mental activity. We could see here evidence of why focusing on the breath helped a person escape the House of Habit—just immersing oneself in the omnipresent sensation of breathing was enough to pull energy away from rumination, judgment, and expectation.

Here we had the brain equivalent of the escape scenes that anchor *Easy Rider*, *Thelma and Louise*, and *Mad Men*—sense foraging into the breath provided an escape, an opportunity to leave the fraught world of ideas behind. Yet the questions remained: What do experts do to make such experiences more than just a flight into the sensory wilderness? How do experts use their time in the wild to deal with the realization that after the relief, there's still laundry to fold?

To assess such expertise, we asked our participants to fill out a questionnaire known as the Multidimensional Assessment of Interoceptive Awareness (MAIA). Responses to this survey indicate how available and trustworthy a person's own body sensations feel to them. For example, someone who has a high MAIA score will agree with phrases such as "When I am upset, I take time to explore how my body feels" and "I can refocus my attention from thinking to sensing my body." Higher scores are linked to greater mental health, and scores tend to be higher following body-awareness training such as mindfulness meditation, yoga, and physiotherapy.[8]

We looked at how variation in MAIA scores predicted the brain shutdown effect we'd observed during breath tracking. We found something surprising: while most of the cerebral cortex still shut down in high MAIA scorers, the brain's salience network did not. Everyone seemed able to disengage from habitual thinking patterns, but those who were the most comfortable with sense foraging maintained salience network activity that allowed them to appreciate new sensations in this quieter mental space.

Our brains are almost always processing sensory information, but usually without our conscious awareness. The DMN guarantees that there are many other processes running—our thoughts, our external senses, our plans and memories and worries for the future, and all of these contribute to cortical noise that is often louder than the subtle but ubiquitous representation of raw sensation. Learning to sense forage is a critical skill, but we can become victims of our own success, shutting off our ability to make new meanings in our efforts to shut out the noise. The type of model updating that empowers taking action requires moving beyond the novelty afforded by sense foraging, to avoid getting lost in the rapture and relief of entering the unknown.

In chapter 7, we introduced three steps leading into sense foraging: recognizing when you're stuck, setting an intention to sense forage, and then doing it. To toggle *back* from sense foraging, here are three more.

STEP 4: LABEL WHAT YOU FIND

In all walks of life, there are people who are highly attuned to sensation. With one touch, a skilled massage therapist can get a better sense of your body's tension, and exactly where you're carrying it, than you have yourself. The skilled chef can take ordinary ingredients and combine them in ways the rest of us would never imagine. A statistician can see a collection of dots on a chart and detect patterns of correlation the rest of us miss (try it yourself by going to guessthecorrelation.com). A psychotherapist can catch subtle changes in a client's tone or detect topics being avoided amid a rambling narrative.

Yet what makes each of these experts valuable is not simply their ability to tap into the sensory world but their ability to toggle back and apply what they've found to the problem at hand. Through sense foraging, professionals develop generally accepted best practices. In formalizing this process, each field develops a vocabulary that allows professionals to articulate fine-grained sensory distinctions. The sommelier can point out subtle undercurrents of flavor that allow her to pair a certain wine with a certain meal. A dancer knows how to translate a rich vocabulary of felt movement to express a narrative or emotion to their audience.

To be useful, though, this kind of enhanced vocabulary of sensation needs to connect foraging to actionable knowledge. The first step in creating this link is to *label* our experiences. This naming reduces ambiguity as we dive into the chaotic world of sense foraging. It demands that we make specific distinctions amid the blur of sensation.

Interpreting facial expressions is one obvious example in which labeling comes into play. Prior research suggests that without labeling, meanings are often determined for us by the DMN. For example, emotionally expressive faces, such as one showing an angry grimace or wide-eyed fear, tend to activate the amygdala, a component of the salience network that serves as the brain's alarm signal. These expressions demand our attention, eliciting alarm without a clear solution, which puts the brain at a crossroads. Without further intervention, the salience network activation will likely trigger a habitual response driven by the DMN. When we see an agitated friend, we might immediately move into a reassuring, problem-solving mode, wanting to find out what caused the upset and how to address it. Yet the alarm signal can also push us toward exploration and conscious deliberation, supported by the dorsal attention network. A common tip for improving relationships when our friends are upset is to just listen and try to understand, rather than jumping in to try to solve the problem. Foraging can reveal emotion, and when it does, the question becomes, Are you prepared to learn something new and respond in a new way?

To counter the DMN's tendency to supply us with automatic meanings, Matthew Lieberman and colleagues examined whether labeling different features of a face determines how that face is processed by the brain.[9] Lieberman's research team discovered a formula for transforming how we react to emotional information, letting go of attempts to fix while exploring. The key was to explicitly name the emotion. When participants were asked to ignore the emotional information in faces and instead label gender, the salience network would continue to fire its alarm signal in the brain. On the other hand, when they were instructed to label which emotion the face was conveying, it seemed to hit the snooze button, letting the salience network deactivate once more.[10]

Further studies demonstrated that people who valued sense foraging (on a mindfulness survey) showed a much greater salience network snooze effect when labeling emotional expressions. They also activated frontal brain regions for conceptual reasoning. By parsing salient sensory information into language, we reduce the urgency to respond while maintaining the gist of what's happening. So, the conceptual system doesn't have to oppose feeling in the moment—instead, *labeling* helps us to better understand emotion, just as a good friend tries to understand and validate our experience rather than jumping to problem-solving or changing the subject.

Toggling from sensation into language is a concrete act for integrating the proceeds of sense foraging into an updated model of the world. A sensory label goes a long way to resist the DMN's tendency to construct perception based on expectation rather than direct experience. Unexpected sensations *always* prompt the brain to respond and make sense of the situation. Like a barking dog as it defends its home, the salience network is not annoying us on purpose. It's waiting on reinforcements, for our conscious minds to give the signal that we're alert and on the case. Whether we respond creatively or by lapsing into habit is not really its concern; what matters is just that we respond. Labeling means making a choice to treat the salience network's "barking" as important and worthy of investigation, rather than some blip on life's radar to be smoothed over.

Once you have learned the benefits of sense foraging, labeling represents a commitment to fully witnessing your experience without jumping ahead to a story about it. It is tempting to think this witnessing means that you fully understand an experience and how to control it (active inference), but that would be premature. What we mean here is to do sense foraging as it is described in chapter 4's exercise, Foraging for Change in Sensation. Our goal is to see the many changing parts that make up an unexpected feeling and to become familiar with them. Every feeling has its own set of sensory dynamics, intensities, and locations in the body, which allow us to explore at the level of sensation even if our habit is to jump into a story about what it means to feel this way. *The labeling we are describing here is about using language to capture these sensory qualities, not their meanings.* Here's an example to make things more concrete.

Kristina felt like something was wrong when she arrived home from work. She had this unsettled feeling in the pit of her stomach that she only really noticed when she arrived at the quiet of her apartment. Rather than pour a glass of rosé, Kristina decided to take a moment to do some sense foraging. She decided to just sit for a while and let her experience unfold, foraging deeper into the feeling that something was off, delving into the murky sensations in her stomach that were running beneath the surface. At first it just felt sort of bad in her stomach, but over a few minutes, she noticed other details, almost like an ebb and flow of energy, making the feeling more intense at one moment, more subtle the next. She was able to locate the sensations in her lower abdomen, a tightness in the belly, a dull ache rather than a sharp pain, a muddy warmth rather than coolness. Slowly, Kristina got to know the feeling. Without knowing why she was feeling this way, or what it meant, Kristina was able at least to recognize when it was present or absent. She decided to label this feeling as "gray" and resolved to watch out for it in the future.

Labeling is the act of recognizing experience by *naming* it. It draws us closer to an experience by helping us stay put at the level of sensation rather than avoiding or moving past it. Kristina's label was arbitrary, but it created an identity for that feeling in the pit of her stomach. She could use the label to stay with sensations and buy herself some time to explore, rather than succumb to pressure to explain and control the sensation (an active inference) when she really had no idea what was going on. Labeling is a type of active inference, but it is one that asserts that it is sufficient for now just to recognize, not to explain or fix.

STEP 5: GIVE YOURSELF OPTIONS

What we normally do with feelings is incorporate them into our internal narrative, driven by a need to immediately make sense of the unexpected and determine what needs to be done. Yet while it is satisfying to explain away a surprising feeling, it also robs us of the opportunity to consider alternative viewpoints. A sensory label interferes with our habit of immediately jumping from sensation to judgment or problem-solving. In neural terms, it prevents the DMN from just smoothing over the complexity of a surprising feeling with a habitual interpretation. Labeling a feeling means acknowledging that our models *can't* explain the current situation and not trying to force it. While this might be a bit unsettling, from our perspective, that's a good thing. Labeling instead of interpreting holds the door open for future insight. If the feeling never recurs, then no harm done, but if the feeling comes up again, our greater familiarity with it can come in handy. A label helps us quickly recognize a new feeling and explore how it might be related to things going on around us. We may discover surprising and novel interpretations for what the feeling is really about.

Having taken some time to explore and label her experience, Kristina went on with her evening. Over the next few days, life unfolded as usual, although Kristina kept an eye out for that "gray" feeling. It didn't show up when she dragged herself out of bed in the morning; it didn't show up on her commute; it didn't even show up when her boss gave her coworker credit for one of her ideas. And then as she walked past the park on the way home, she noticed children playing, dogs barking—and then, yes, there it was. As she watched the dogs running around, thrilled to be out in the park, the feeling was there—"gray." "Huh," she thought, "that's weird." She took a moment to forage into sensation and confirmed that it was pretty much the same feeling she'd had a few days earlier. She continued home.

Over the next few weeks, Kristina tried to note where she was and what she was doing when "gray" showed up. It turned out not to be related to spicy food at lunch or having a few drinks the night before. Kristina noticed that "gray" tended to occur most often when she saw dogs playing with their owners. And it made sense in a way. Kristina's prized companion, a wire fox terrier named Alfie, had passed away a few months earlier. Kristina had grieved at the time but thought she was over it. The "gray" feeling in the pit of her stomach revealed to her that she wasn't as far along as she'd thought—maybe she still missed Alfie.

STEP 6: CHOOSE WISELY

In the end, you will probably have to make an active inference around sensation, whether or not you spend time sense foraging first. We are wired to solve problems, and an unexpected feeling or event presents a problem to solve. The question before us is whether we are solving the *right* problem. Imagine if Kristina, in the absence of sense foraging, went online and looked up "stomach pain." Immediately she would see possible medical problems, such as ulcers, irritable bowel syndrome, gastritis, diverticulitis, and even stomach cancer. All important problems to solve, but all missing the true cause. Think of the worry, the time, the distress, and ultimately the frustration after trying to solve this particular problem by only considering medical options. Even if she can do a quick analysis and rule out many of the possibilities, it still takes time, effort, and worry, and in the end it won't reveal an unexpected cause such as grief.

Toggling comes from an understanding that we are wired to solve problems but that we need space to consider which problems to solve before we act. Sense foraging provides us with this space, and perceptual updating allows us to discover unexpected causes. We may feel that we need to do something active in the face of an unexpected feeling, but the first active inference can be to sense forage rather than explain and fix. Anchoring feelings with sensory rather than conceptual labels allows us to detect patterns in our experience rather than forcing a familiar narrative.

Kristina's "gray" label gave her a way of staying put at the level of sensation when the feeling showed up again. This allowed her to explore other sensory features of her experience when "gray" was present, rather than rehearsing a narrative straight from medical websites. Anchoring her attention at the level of sensation, she was able to notice an unexpected context linked to her feeling (the dog park). When Kristina considered "gray" again, she now had multiple interpretative options. What's more, these options were not all equal. Some came from what one would expect from the outside looking in, in other words, medical opinions, but the possibility that she might be grieving for a pet was uncovered only by looking from the inside out. Identifying a familiar feeling while staying engaged with sensation allowed her to discover the underlying context.

In the end, Kristina could still interpret "gray" however she wanted. But we suggest there is an important choice to be made when navigating our lives: do you want to pay more attention to a map or to the road on which you are actually driving? A map is something created by experts to describe what usually happens between points A and B, but no map is going to let you know about unexpected bumps, detours, or hazards. Keeping your eyes on the road is just as important as having a destination in mind. There isn't any wizardry or secret calculus to choosing wisely among options. Wisdom lies in creating the conditions in which you don't blindly follow the map of expectations without keeping an eye on the road ahead.

Several years ago, we were in Amsterdam for a conference to present the results of our research on sense foraging and depression vulnerability. We had a moment to chat with Ajahn Amaro, the abbot of the Amaravati Buddhist Monastery, about the relationship between meditation practice and its benefits. He offered an analogy based on his own extensive experience in teaching meditation: "Meditation is like maintaining your car," he said. "But what good is all that careful maintenance if you're driving in the wrong direction?"

What Ajahn Amaro was telling us was that having the right understanding of when to use any meditation practice is more important than rigid adherence to any particular practice. A meditation that promotes calm might be wonderful in times where one needs to endure an unavoidable stressor, but it robs a person of the energy to act when quick intervention is needed. Seeing a child running into a busy road is not the time to practice equanimity, even though it might be the right practice most of the time. What's important is taking the time to investigate what the problem really is before attempting to solve it. Somewhat ironically, we must first toggle away from active inference into perceptual inference before we can act effectively.

Exercise: Don't Cry Over Spilled Cabernet

The DMN can easily map out a solution to the unexpected rather than allowing us to explore the situation first before selecting a strategy. This is especially true when situations affect our reputation or self-esteem. The default map often focuses attention on protecting the ego, with rumination and worry over how we are perceived by others leading to attempts to control perception rather than address the upset itself. To see the effect, take a moment and try to place yourself in the following scenario:

You've been looking forward to this evening for a couple of weeks—a celebration at your favorite bar in honor of Jennifer's promotion to regional manager. There'll be a nice mix of friends and coworkers along with a few games of darts to round things off. At the table, everyone seems to be talking, the servers are bringing trays of appetizers, and people keep stopping by to congratulate the guest of honor. You're seated across from her, and in the middle of this commotion, you stretch your hand out to get a glass of water and inadvertently brush a wineglass next to it. You watch in seeming slow motion as the wine spills onto the table. You try to alert Jennifer to the red sea of cabernet coming her way, but she reacts too slowly to avoid a few drops dripping onto her dress.

As you realize what you've done:

- Think about what consequences this has for you and how you see yourself.
- Think about how guilty you feel.
- Think about whether you are clumsier than you thought.
- Think about whether you've ruined everyone's evening.

Where does all this *thinking* lead? Perhaps you spend most of the evening reminding people about the event and making self-deprecating comments, hoping to elicit sympathy. When you go to bed that night, you're still going over the scenario in your mind with a measure of regret. What's more, you feel apprehensive the next time you're invited to a dinner party, because you never forgave yourself for what happened.

In the preceding response, you've followed the well-trodden path, where judgment and self-recrimination focus you on solving a reputational crisis, but is this really the right problem to solve? You have another option. Take a moment to place yourself in the scenario again. You are in the same restaurant with Jennifer, coworkers, and friends. Your hand brushes the wineglass and it spills, splattering Jennifer's dress with wine stains. This time, though, instead of being focused on yourself, your attention stays squarely in the moment, able to fully engage the senses, which allows you to label the experience quite differently.

After spilling the wine, you *toggle* into sensation and label your experience:

- You have spilled something.
- You feel flushed, blood rushing to your face.
- You notice everyone looking at you.
- You label yourself as "embarrassed."
- You explore what your options might be rather than try to fix your embarrassment.
- You look around the room and take in what's happening.
- You notice the wine is still spreading along the table.
- You notice the people around you aren't angry, just surprised and even a bit amused.
- You *toggle back* and judge between apologizing and cleaning up the wine that is still spreading.
- You *choose* to help mop up the wine and joke about the situation.
- You make eye contact with Jennifer, smile apologetically, and ask her to please send you the bill for dry-cleaning her dress.
- With the spill mopped up, you realize that further apologies are not needed.

By toggling into sense foraging in the face of the unexpected, you accomplish two things. First, perceptual inference gives you a reprieve from the habit of seeing everything as a reflection on you, following the map for protecting the ego: repeatedly apologizing and making the spill the focus of the evening. Today's youth have a name for this tendency—"main character syndrome"—and it is not an endearing habit. Freed from self-obsession, you observe that other people don't care about the spill all that much and that the actual problem is stopping the spill from spreading. This realization empowers you to toggle back into active inference. You can be there for others instead of turning inside yourself, and instead of worrying about how this event might make you look bad, you can now focus on getting the evening back on track. By toggling between your sensory awareness of what's really happening—also known as "reading the room"—you find an emotionally attuned way to address the spill and also an emotionally intelligent way to let go and enjoy the evening.

REPEAT AS NEEDED

It may be easy to look at sense foraging and pluck out one element that you enjoy, embracing it without seeing the full process. Yet as illustrated previously, foraging is part of a larger cycle of balancing exploration against reliance on existing knowledge. *Toggling* is seeking this balance rather than escaping into sense foraging. Not all existing knowledge is problematic and in need of updating. Sometimes, our default maps might be perfectly adequate, while other times they lead us astray. Repeatedly toggling back and forth between sensation and meaning making allows you to get a better handle on new ways of seeing the world, even if it takes a while.

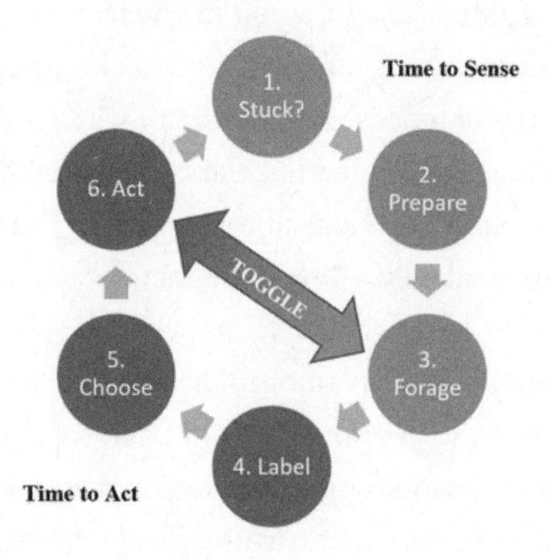

Psychotherapy, since Freud's time, has promoted the popular notion of insight, a powerful moment of emotional clarity and understanding. And we agree that insight is an important product of sense foraging, as discussed back in chapter 4. But anyone who's worked in the field of mental health will tell you that the eureka moment of insight doesn't last. What does last is the much more tedious and decidedly unglamorous process of working through personal experience. This means continuing to reflect on the meaning of experience well after the glow of momentary insight has faded. If sense foraging, toggling, and model updating are all steps on a feedback loop, then it is important to consider where you find yourself in this cycle. Is it time to sense or time to act?

The binary choice at the heart of each moment of surprise or discomfort is familiar to millions of people. It is expressed perhaps most clearly in the Serenity Prayer, written by the American theologian Reinhold Niebuhr, which is a staple of twelve-step recovery programs:

God, grant us the serenity to accept the things we cannot change,
the courage to change the things we can,
and the wisdom to know the difference.

Serenity comes from finding a way to maximize both agency and presence in any given moment. Agency: to change the things we can. Presence: to accept the state of the world when we cannot change it. And wisdom: the insight, intuition, or life experience to realize that these are always the two choices we have in the face of upset, and that we'd better choose wisely or we'll continue to suffer from the mismatch. This is the essence of what we mean by toggling.

Of course, it's easier said than done to accrue the wisdom part. And it's interesting to consider whether we're really brought up to take both options seriously. Some of us were taught that agency is the name of the game, that a successful life involves grabbing problems and wrestling them to the ground. From that perspective, any passivity is weakness—followers of the agency-or-bust school of thinking are exemplified by *South Park*'s Eric Cartman, who famously opines, "Whatever, I do what I want!" Or perhaps we've been indoctrinated to think we're meant to be passive vehicles through life, as some more traditional societies insist—especially for women—and that acceptance is a virtue even when we're being suppressed or harmed. The polar opposite of Cartman, Charlie Brown's blanket-toting friend Linus passively endures ridicule by his domineering older sister, Lucy, claiming that "learning to ignore things is one of the great paths to inner peace."

The Serenity Prayer is a reminder that, rather than falling for either extreme, we always have a choice when our expectations don't align with our experience. The motivation comes from the way our brains work, but we need the wisdom to know what to do with it. What is exceptional about this prayer is the way it acknowledges that it's the disconnect between prediction and sensation that drives our discomfort, and as such it may be addressed by tackling either horn of this dilemma. Either my expectations or my reality must change, but which one?

It's important to point out that each mode of inference has its dark side, which can easily feed its own brand of resentment. Active inference leads to a sense of agency when it accomplishes its goal, but when we don't succeed, it can make us feel even more powerless. Sometimes we descend into what psychologists call *learned helplessness*, a gateway to depression. Perceptual inference is the courage to accept *what is* over *what we wish for*, but relinquishing our deeply held desires might narrow our sense of self, inviting further disconnection and isolation. We may end up resenting no longer fitting into a world that once had a place for us.

So, what is wisdom for each of us within this paradigm? This question is at the heart of self-knowledge, purpose, and meaning in life. It may well be that the feeling of "you" is little more than your history of reconciling between active and perceptual inference as you learn to make your way through the world, changing what you can and adapting to what you can't.

SENSE FORAGING AT WORK

Taking one more look back at our friend Shanice, we can see that, once she made up her mind to leave PetSmart, she followed the young entrepreneur playbook, and everything seemed to fall into place. But her playbook may have been missing a chapter on how the manic intensity necessary for launching a business may not be suited to ongoing management over the long haul. If Shanice had been able to relate differently to the self-recriminations about not being able to keep up the pace, she might not have driven herself into a burned-out state of exhaustion.

Fortunately, her crying jag in the stairwell opened her up to being more receptive to some of her coworkers' feedback, and she decided to pause and take stock. She realized that the feelings of hopelessness and burnout were a new development. In the early days of her startup, even when she'd be tired from working killer hours, there was mostly a sense of satisfaction after a day of giving her all. But now that good feeling had given way to a sense of dread and fatalism in the face of a commitment that seemed overwhelming. Resignation and anxiety had replaced her hopeful energy.

Shanice recognized that she was stuck, trapped in a pattern of exploiting her existing knowledge without exploring other possibilities for how to accomplish her goals. Her DMN, triggered by a sense of threat, fueled further attempts to solve problems the same old way. At the same time, stress was shutting down her ability to be creative and discover other options.

Shanice's mom and her best friend had dropped hints about the importance of making time for restorative and relaxing activities, but she routinely dismissed these as frivolous. Who could afford to devote half a day to go to a spa, sit by a lake, or wander around a hiking trail? Too little time and too much to do.

Even though she was slightly irritated, Shanice knew their advice was coming from a good place. She had to admit that much of what they saw was real—the weight loss, the exhaustion, how she constantly second-guessed herself. In fact, these issues had become so familiar to Shanice that she had a name for them: weariness. She recognized weariness when she had trouble keeping her eyes open in the afternoon or sat immobile, staring out a café window thinking about her next move. "There's too much weariness in my life and it's not going away anytime soon," she thought to herself.

Letting some of her mom's advice sway her, Shanice decided to pep things up a bit and reached out to Marsha, an old friend from her business school days, and asked if she would like to take in a movie. Marsha was delighted and they made plans for the following Friday night. Shanice arrived at the theater and waited around, but Marsha didn't show up. After calling her and with time running out, Shanice decided to go into the theater anyway; Marsha could always text her when she arrived. At first, sitting in the dark and watching the film helped to take Shanice's mind off her worries, but soon she found herself thinking, "I shouldn't have to go through this alone; there should be someone here at my side." And there it was again—weariness, her body slack in the seat, hands fidgeting. But something was different. Shanice had made a connection between weariness and seeing herself as alone and without support. For the first time, finding someone she could trust and share her struggles with made sense as an action she could take to relieve this unwanted and unwelcome feeling.

That evening, after coming home from the movie, Shanice made the decision to hire an operations manager. The tyranny of thinking that she had to bear the burden herself simply faded away, and suddenly she was open to letting someone else take over the daily decision-making around packing, shipping, inventory control, and accounting. She would narrow her immediate focus to sales and growing the business—the things she did best.

While hiring and training a manager was nerve-racking, the effort was ultimately a huge success. Far less exhausted, and feeling more like her old optimistic self, Shanice could still exploit her amazing ability to grind it out when she needed to, but otherwise, she reserved time and space to explore the beauty of the world around her. Toggling back and forth between her hard-driving business persona and her growing sense-foraging practice allowed her to stay connected to a larger sense of self, not as a hamster running on a wheel, but as a fully engaged citizen of the sensory world.

PARENTING: SOME THINGS ARE TOO IMPORTANT NOT TO CHANGE

We have discussed accepting a situation rather than fighting for change, but toggling is not only for achieving acceptance. Sometimes, after toggling into sense foraging, what becomes clear is that our inclination to act based on existing knowledge still seems correct; it was only our uncertainty that prevented us from giving ourselves fully to the solution. Sometimes we can get stuck thinking we have to accept something when change is needed—the other horn of the Serenity Prayer. Such situations are most likely to arise when something important to us is at stake, and we have more to lose by not acting than by taking a risk and seeking to change the world. Consider the following example.

"Am I a hypocrite?" As an executive business coach, Javier preached the benefits of self-reflection for a living. Just the week before, he had emphasized to a client how awareness puts a wedge between stimulus and response. All you really need is to pay attention and accept what you see.

And yet here he was, again. Vein in neck bulging, exasperation in his voice as he roared at the kids. They weren't putting their shoes on and were going to be late for school. It didn't matter if they got ready early; the dawdling seemed to continue until Javier lost his temper. It made mornings stressful for everyone.

"Great job, problem-solving expert. Shouldn't perceptual inference do the trick? After all, you prescribe it to all your clients." Javier told himself he should just *accept* lateness. His children's ability to self-organize was something best left to the whims of fate and surely beyond his control.

Javier resolved to do some foraging. For several weeks, he set early-morning intentions in the bathroom, using the cold tiles against his feet as a reminder. He vowed to witness himself as he struggled to get his kids out the door each morning. To allow perceptual inference to unfold.

Yet most mornings he found himself exasperated and frustrated, caught between the horns of being an ineffectual parent and being an overbearing tyrant.

Then one day Javier *got* it: he updated his model, but not in the way he expected. Perhaps that's the point. Javier realized that he didn't *want* to be okay with his kids being late for school. At the same time, he understood that getting frustrated and angry was his model for dealing with lateness, and that model wasn't working for him. He became ready to change—not by accepting lateness, but by accepting that his model wasn't working. He was stuck.

So, Javier toggled. He had seen the problem clearly; it was time for action.

He started coming up with different ideas each morning to get his kids out the door—the only criterion was that each idea had to be *fun*. For example, one morning he burst into the kitchen, warning the children that the front of the house was besieged by hungry beavers (we're in Canada, eh?), and only a rapid escape out the back door could save them. On another occasion, he transformed into a fearsome beast, scavenging the halls for shoeless children and showering them in tickles.

It took work, but it worked. Slowly, getting out the door became less a point of contention and more an opportunity for improvisation and play.

Exercise: Commitment to Change

If you practice sense foraging, you will eventually notice two things. First, you'll become familiar with being stuck in patterns that are running you. Second, you'll decide that at least one of these patterns is something that you want to change. For instance, you may be dissatisfied with your weight, but by hook or by crook you find yourself eating dessert each evening. You may wish you were less judgmental but find it irresistible to gossip about your coworkers. You may bemoan your sedentary lifestyle but find yourself zoning out to the television each evening. In each case, you can see a model for self-care that needs updating, even if you aren't sure what to update it with.

There is a moment when sensing and labeling the familiar can be met with a yes or a no. This is the moment of toggling from sensation back into evaluation. Are you okay with the pattern continuing?

If the answer is yes, you can leave the pattern alone and toggle back to foraging for new sensations and possible patterns. But if it is a no, your foraging needs to take a different turn—now you are foraging for a better model, not just an understanding of the existing ones.

The following exercise is for when you've seen a pattern and realize change is needed. In the space provided (or, if you prefer, in a notebook) write down and respond to the following prompts:

My familiar, unwanted pattern is:

____

____

What I have tried that *didn't* work:

____

____

What I haven't tried yet that *could* work:

____

____

Pick one of the untried strategies and commit to exploring it until either it helps address your issue or it doesn't; if it doesn't, you can add it to the "didn't work" list. If change is important to you, get back to the untried list and try the next thing. Eventually, you will find it. Channel your inner Javier! Once you are clear on what you are foraging for, you just need the courage to keep trying until you find it.

10

A SENSIBLE FUTURE

The doctor of the future will give no medication, but will interest his patients in the care of the human frame, diet, and in the cause and prevention of disease.

—THOMAS A. EDISON

FROM PATIENTS TO PARTNERS

The COVID pandemic fell like a tree branch through the House of Habit's roof. As the arbitrariness of workplace routines was laid bare, billions of DMNs set off a global tsunami of prediction errors, and business as usual simply stopped making sense. As pandemic restrictions relaxed, employers faced an unprecedented change in employee expectations. Having tasted a change from the status quo, many resisted the call to fall back into default workplace models after navigating the challenges of learning to work from home.

The perceptual inference that life *could* be different empowered the active inference that it *should* be different in the form of the Great Resignation. For the first time in a long time, there were far more open positions than job applicants, leading to an incredibly competitive hiring environment. Savvy companies realized that the employment model had changed, responding by introducing nontraditional incentives. While corporate health plans historically reimbursed medically necessary expenses, like glasses and knee braces, companies like Zappos, Google, and Intuit now pay for you to take music lessons, join a gym, or learn to cook.[1] Such benefits may speak well of a company's culture but also raise the question, Why would insurers, usually a wary, fiscally conservative group, fund elective, recreational activities?

Part of the answer is captured in Thomas Edison's prognostication at the beginning of this chapter. The pandemic seems to have catalyzed an increase in our expectations for self-care and risk management. Actuaries have realized it's cheaper to promote employee health by giving people *access points* to nip their own problems in the bud, leading to reduced medical costs down the road and to a healthier bottom line, for both you and them. This change is not simply due to the COVID-19 pandemic but has been building for decades. Individual responsibility for maintaining health has infiltrated medical language, with providers ceasing to refer to their charges as "patients," literally waiting for a doctor to cure their suffering, and referring to them instead as "clients," partners in the enterprise of self-care.

Giving people the agency to look after themselves opens the door to promoting health rather than only managing illness. A new generation of biohacking technology, such as smartwatches and rings that track heart rate and activity levels, allows us to detect risk markers before they manifest as clinical disorders. But it wasn't always so. Healthcare has historically been interventionist, in which a doctor responds to acute medical conditions but shies away from health promotion among the well. A century ago, a common quip used to disparage physicians exploring preventive medicine was "How can anyone make money taking care of healthy people?"

The assumption that people are generally too healthy to benefit from medicine turned out to be ill-founded. Following World War II, survey after survey indicated that people—at least Americans—weren't all that healthy. The advent of standardized high school fitness tests (toe touching, sit-ups, and leg lifts) revealed a shocking disparity between the average American youth and their European counterparts. A whopping 56 percent of American high school students failed the test, while only 8 percent of European children did.[2]

The fitness movement in the seventies and eighties changed all that. Led by innovators such as Ken Cooper, the Western world underwent a radical binding of personal agency and illness prevention that was then scaled worldwide. Ken gave people a motivational hook into exercising, pioneering a set of exercises known as "aerobics," what we would now call cardiovascular exercise. He preached preventing heart attacks by prescribing treadmill-based assessments and exercise programs. While this approach is now standard practice in cardiology, in the mid-sixties it was seen as dangerous and misguided, adding stress to bodies already near their breaking point. In fact, Cooper was once hauled in front of his state medical disciplinary committee for complaints that he was endangering heart disease patients by having them run on treadmills.

Over time, however, clinical research supported Cooper's claims that an ounce of cardiovascular stress was worth a pound of bypass surgery, making clear that one ought not to wait for chest pains before considering dietary and exercise habits.[3] Central to Cooper's message was that *fitness could and should be improved by training.* Yet herein lay a challenge that our neuroanatomical tour could explain: a nation of DMNs had a "mechanic" model of health. Other than the occasional tune-up, you wait for something to go wrong before you take your car into the shop, looking under the hood being a job best reserved for professionals or enthusiasts. Cooper wanted to update these models to help the public see their health through a gardening metaphor instead: healthy plants require engagement and work throughout the life span, with each gardener ultimately responsible for the state of their lawn and flower bed. Yet to somehow update millions of models to see the possibility of personal maintenance or even growth, *access points* for sense foraging into the emerging fitness space were needed.

Cooper's preferred training regimen was running, which became a huge trend in the seventies. Tennis, yoga, weight lifting, and aerobics classes also saw huge upticks in participation, alongside an independent market for self-help books and televised classes. The fitness boom handed agency over to the individual and put the means of improving health in their hands. By providing multiple access points and a marketplace willing to sell them—Jane Fonda's workout videos or Richard Simmons's television shows—fitness instruction was scaled up for the masses. As people foraged and found empowerment in exercise, there was a parallel revolution in other health behaviors, from quitting smoking to trading in martinis for wheatgrass smoothies. Together, the combination of access points and role models fueled the belief that pursuing a fit lifestyle was both good for you and achievable. Decades later, many of us have internalized the notion that we are responsible for our physical health. We are taught through a combination of nutrition, physical education, and sex education programs that we must learn to act as constant gardeners rather than hapless motorists.

Today, we're on the cusp of a similar transformation in mental health. More and more people seem to be losing faith that conventional medicine can cure their depression or anxiety and are yearning for better options. As countless DMNs throw prediction errors between the expectation of happiness and the reality of languishing, it feels like we're at the precipice of a new *technology of the self*—a term popularized by the French philosopher Michel Foucault.[4]

Foucault advocated the use of such simple "technologies" as the diary, which allows someone to forage through their own mental habits. The premise is simple: judging our past selves is not as threatening as judging the present self. Who among us has not had a few choice words for the overconsumption and procrastination of their past self? When you read what you wrote a week ago, you get a glimpse into "past you," and "present you" can easily forage through this ancient figure's musings without taking too much offense. In other words, reviewing one's own diary promotes *decentering*, one of the central mechanisms of sense foraging. In turn, this practice can yield insights that motivate reflection and change, providing greater agency and presence in building the life you want.

Sounds good, no? And yet, it has been decades since Foucault's suggestion. If diaries alone were going to transform mental health, we should have seen the fruits of people's labors by now. The yearning for change is here; what are we still missing?

IMPORTANT FAILURES

Foucault is hardly the only one to opine on wellness routines that have fallen short. For example, there are striking parallels between the fitness boom and the recent popularity of mindfulness meditation. Both movements caught the public's imagination, were promoted widely, required that people step outside their everyday routines, and promised tangible health benefits achieved non-pharmacologically. In fact, the evidence supporting mindfulness's benefits was so compelling, and the need so great, that the British government supported a huge trial of bringing mindfulness into schools. Mindfulness was pitched as a universal program for teaching youth social and emotional skills that would protect mental health.[5] Investigators worked with local school systems and succeeded in embedding ten forty-minute classes into their curriculum. The My Resilience in Adolescence (MYRIAD) study took eight years to complete and involved one hundred researchers working with twenty-eight thousand adolescents and 650 teachers in 85 schools in the United Kingdom.

After all this effort and innovation, students receiving school-based mindfulness training were no better off than those without the training. Some students even showed a deterioration in mental health outcomes. Flummoxed, the investigators dug a little deeper and found that very few students practiced the skills they were taught. Those who did practice received the hoped-for benefits, but they were a tiny minority.

Other large-scale, school-based studies showed similarly disappointing results from online CBT, as well as the practice of "nudging" with snappy reminders.[6] The problem was that each of these attempts to promote mental health tried to find the *one* right thing to do, rather than providing multiple access points. Students in MYRIAD made this loud and clear in surveys. They appreciated that someone was taking enough of an interest in their mental health to put MYRIAD in the curriculum, but they found forced meditation boring, and they would have preferred to have had the free time to engage in activities they cared about.[7]

The lack of options in the MYRIAD study to support mental health is comparable to the state of American fitness in the 1950s, as if the government had mandated that everyone do push-ups. No other programs, no public relations campaign. Just push-ups. Fortunately, advocates formed the President's Citizens Advisory Committee on the Fitness of American Youth, which funded a massive PR campaign.[8] And they were never dogmatic about what form exercise should take. Over time, the understanding grew that a sedentary lifestyle was a problem, and that fitness was a solution. By the 1980s, gyms, health clubs, 10K charity races, and bicycle teams were everywhere. Championing the culture of fitness, rather than any one access point, had made the movement a success.

Today, if you believe that getting stuck in unhelpful patterns is a problem, and that toggling into sense foraging can be a solution, we believe there's an access point that will work for you. You might have to engage in a bit of foraging, but you will know it when you find it.

WHAT DOES SUCCESS LOOK LIKE?

> You become what you give your attention to. . . . If you yourself don't choose what thoughts and images you expose yourself to, someone else will . . . and their motives may not be the highest.
>
> —EPICTETUS

Once you find an access point that works for you, what kind of changes can you expect? That depends in large part on the kind of changes you want.

A recent study on mindfulness training made a splash when the popular press interpreted it as having shown that *meditation makes people more selfish.* Yet behind the clickbait title was a more nuanced story. Following the meditation, participants had a chance to stuff envelopes to help a charity campaign raise money to combat homelessness. But instead of meditation making everyone more willing to donate their time, the effects of meditation depended upon people's preexisting attitudes.[9] Meditators who described themselves more through their relationship to others (I am a daughter, I am a caretaker) stuffed 17 percent more envelopes than a control group. However, meditators who described themselves more through personal characteristics (I am determined, I am resourceful) did the opposite—they stuffed 15 percent *fewer* envelopes.

This study rattled some DMNs, violating the cliché that meditation universally drives selfless compassion and prosociality. Yet in their zealous defense of the mindfulness movement, the study's critics made an error in assuming that everyone *wanted* to get the same thing out of mindfulness training. The data were clear: meditation "success" meant expressing one's internal values, regardless of what the teachers and theorists expected.

From our perspective, this is fine. Empowering individual sense foraging is not the same thing as buying into someone else's value system. Sense foraging *should* lead to a range of different outcomes, according to Epictetus's dictum. Each of us has our own priorities and values, and success is learning to direct our attention toward those things.

Fortunately, we are hardwired to perceive the things that we value most. The brain's salience network is masterful in separating the relevant from the irrelevant, as illustrated by the Tse illusion (named after cognitive neuroscientist Peter Ulric Tse) that follows.[10] Focus on the white dot in the very middle of the image. Now, without moving your eyes from the white dot, try to focus your attention on one of the three large gray circles. If you end up moving your eyes, no worries, just focus again on the white dot, and then again try to pay attention to one of the three big gray circles without taking your eyes away from the white dot.

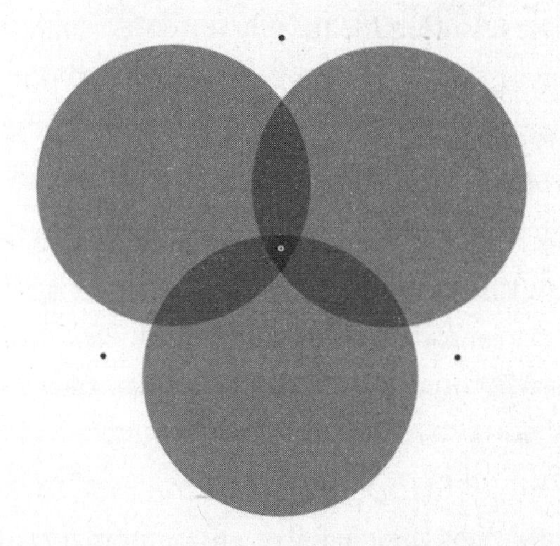

What happens to the circle you're attending to compared to the other two circles? Most people experience the circle they are attending darkening in color, while the two unattended circles fade slightly into the background. No need to move your eyes or change the lighting—merely selecting one thing to attend to foregrounds that thing as relevant, while the mind automatically backgrounds other information that might get in the way. The Tse illusion shows us how our expectations can become self-reinforcing. The DMN provides the brain's salience network with expectations for perception, which juice up perception of the familiar, which in turn validates expectations and strengthens DMN programming.

The implication is that without a specific goal in mind, our default mode of attending leads us deeper into our most familiar ways of perceiving and responding. René Redzepi is a food guy: when he went foraging in the forest, he was likely not paying as much attention to birdsong or animal droppings as he was to potential foodstuffs. Similarly, the things you consider important will be highlighted by your attention, leaving other resources in the background.

Success in sense foraging means getting some control over this process. It means updating your model of what is important or unimportant in any given moment. It doesn't take some big overt action: you need only to *choose* what it is you are going to forage for within the field of your experience. Each time you do, the salience network will foreground any sensations that match your choice; over time, these sensations will become familiar, and the DMN will grow to prioritize your choice in future processing, automatically tagging the rest as irrelevant.

But wait. Doesn't this sound like becoming stuck again? It all depends on whether what has become automatic aligns with your values. The feeling of being stuck occurs when we witness our habits leading us to undesired consequences or blind spots. On the other hand, we don't resent balancing automatically on a bicycle, or learning to tune out a telemarketer asking for just a few moments of our time. We are deeply dependent on habit to get through the complexity of life; habit will always be a part of us, the part of us that desires order and predictability in a chaotic world. The goal is not to get rid of the DMN but to realize that sense foraging gives you the option to weaken and reconfigure it when its programming is revealed to be problematic, precisely by that feeling of being stuck. When the programming is working well, you don't feel stuck—you feel *flow*.

So how do you get to flow? The journey begins by first figuring out what you want to be foraging for. Feeling that you're challenging yourself and growing is an important sign that you're on the right path. But the principle of toggling reminds us to keep asking, "Am I heading in the right direction? Unmoored from habit, am I charting a course toward a life that I value? Or am I foraging to meet someone else's priorities, or perhaps without a clear sense of purpose at all?"

DIFFERENT SITUATIONS CALL FOR DIFFERENT GOALS

When choosing what to forage for, your environment matters. It is hard to find authentic praise for a child who has just smeared oatmeal onto the carpet, or to sparkle with creativity right after a harsh performance review. Finding a way to address these situations in a realistic and constructive fashion seems more appropriate than grasping at unrealistic straws. It also doesn't make sense to forage when our default habits and inclinations are working well—we don't want a world of ambiguity when we see brake lights ahead of us on the highway; we need to get the car slowed down, not to philosophize about the experience of the surprising red light ahead of us.

On the other hand, we *can* get to know the situations that call for sense foraging rather than exploiting existing knowledge. We may feel stuck if we habitually explode with anger when oatmeal is smeared; we may feel hopeless after negative feedback; we may begin to avoid driving for fear of an accident. When we see our programming running us toward undesirable outcomes, the need for sense foraging becomes clear. Yet diving into our senses with no set of intentions seems impossibly chaotic; sense foraging works best when we can make at least a few basic decisions as we enter the process.

The most fundamental of these foraging choices is between relief and insight. Sense foraging can provide *relief* from stress by whisking you out of habit and into the relative freedom of sensation—so deciding to use sense foraging this way is a form of *active inference*, using it to change the state of stress and physiological arousal in our bodies. Foraging for relief can help us regulate our physiology without updating our conceptual models, simply by interfering with the DMN's ongoing defensive habits. This may seem less than ideal; if we are simply escaping from the House of Habit by running into the wilderness, we aren't equipping ourselves to deal with real-world problems in a constructive way. Ideally, we would forage for new meanings, to escape from our preconceptions about the world; here sense foraging empowers *perceptual inference*—updating our conceptual models, which can cause discomfort as we embrace the unknown.

Yet research shows that people have very different ideas about which option is best. For example, we examined how the term "mindfulness" is used by the public, scholars, clinicians, and students.[11] For scholars and clinicians, mindfulness is ideally a way to practice *perceptual inference*. One uses awareness to explore the world, paired with acceptance of what is found to provoke insights that expand the self and one's worldview. But most people aren't thinking about growth and development; they're just trying to get through their day. That's why, for both students and the public at large, mindfulness is a way to practice *active inference* to find relief, disrupting stress-provoking mental habits without much intention to learn something new.

This distinction applies to the casino as easily as to the meditation hall. In one study, participants who staked fifty dollars to play virtual slot machines could choose to escape into the comforting monotony of absorption with their current machine or forage for a new machine that felt "hot" to receive a payout. Yet researchers observed that what determined a preference for relaxing absorption versus expansive exploration had more to do with the situation than with the machine.[12] Left to their own devices, most players liked to explore different machines, but when taxed by a complex memory task, players favored staying put and not rocking the boat.

You may be a naturally curious person, but if, for example, you are also the caregiver for an elderly parent, it's worth the effort to assess how much of a cognitive load you're working under and then choose wisely from that point. What we choose to toggle toward is powerfully influenced by how much mental space is taken up responding to everyday demands.

So, is it better to forage for relief or for insight? The simple answer is that both approaches are perfectly acceptable. The key for the individual is to be clear about what they're foraging for, without putting themself under the stress of having to "do it right."

Exercise: Putting Yourself on the Map

To help you get a better sense of your scope for foraging, here is a brief self-diagnostic exercise. We've set up a *Cosmo*-style quiz based on a well-established mental health assessment.[13] If you'd like, give it a try:

IN GENERAL, DO YOU:	YES	NO
1. Find that the demands of everyday life get you down?		
2. Feel you are drifting through your life?		
3. Feel disappointed in your life achievements?		
4. Find it difficult to maintain relationships?		
5. Take life one day at a time and not worry about the future?		
6. See yourself as a giving person?		
7. See life as providing ongoing opportunities for change and growth?		
8. Feel confident to express your views and opinions?		
9. Believe you have a sense of direction and purpose in life?		
10. Like most parts of your personality?		
11. See yourself as a lifelong learner?		
12. Feel like you can make a difference and contribute to your community?		

Give yourself 1 point for every item you said no to for statements 1–5 and 1 point for every item you said yes to for statements 6–12. Now take a moment to add up your points and try to locate yourself on the life-satisfaction continuum that follows. If your score is in the 0–4 range, you may find yourself struggling or shutting down when dealing with multiple challenges in your life; if your score is in the 5–8 range, you find that things are

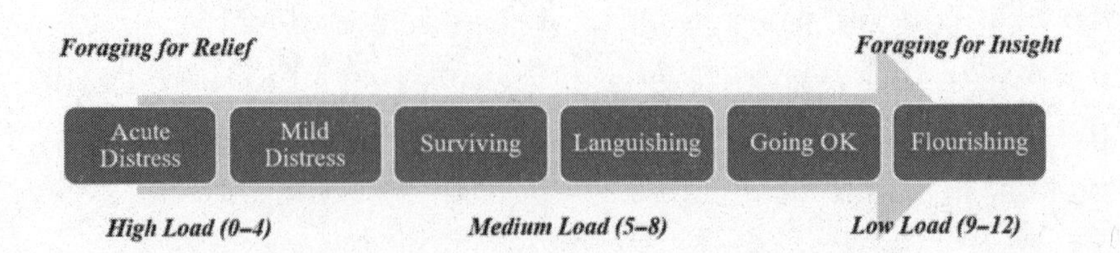

going okay for you; and if your score is in the 9–12 range, then you are approaching optimal functioning.

Your place on the continuum may be helpful in figuring out how heavy a mental load you're carrying. In a high-load situation, the pressure to exploit existing knowledge is going to be intense, and you'll likely feel that relying on habit is all you can do to stay afloat. For example, after even a minor car accident, there are a lot of steps you have to take, such as pulling over, exchanging information with the other driver, taking pictures of the damage, assessing your own physical and mental health, and dealing with whatever plans have now been disrupted. There could be emotions of anger, blame, and disappointment, and possibly even catastrophizing around the financial costs of repairing your car. This might not be the best time to forage for how you can become a socially active leader in municipal politics to make roads safer. There's too much pressure to act now and get through the immediate demands of the situation. Under high levels of stress, the role of foraging changes—you still need to forage for something, but most reasonable is to forage for a sensory refuge from the racing mind. Take a few moments and check in with your body, notice that you are not still in any imminent danger, and allow yourself to calm down and recover the resources you need to act.

Eventually, foraging in the face of stress will provide relief, as the habitual representations of conflict fade into the background when confronted with the raw energy of sensation. This buys you a bit more space—the process of decentering we discussed earlier. It helps move you from resistance mode on the far left of the continuum to becoming curious about experience, somewhere in the middle. It is only when you feel like you have the capacity to be curious and investigate that *insight* is likely to occur. As foraging opens you up to new experience, and you feel less stressed, your objectives for foraging can shift to something larger—gratitude for what you have, joy in daily interactions, donating time to improve the lives of others, you name it. In the absence of pressure to protect yourself, you can aspire to become something more.

You can't control whether you have insight; sometimes you can't even guarantee that you'll experience relief—some days are just like that. But put yourself on the map; be real about how you're doing and, at the very least, choose relief. Don't let yourself get burned out waiting for inspiration when what you really need is a break.

Our research on depression vulnerability shows that even breaking the habit of blocking out sensation when you're stressed takes time. It takes even more time for other habits to loosen up. It's like having a tight muscle that's painful from overuse. The first step is to stop exercising the muscle to exhaustion each day. You might then experiment with stretches and physiotherapy, until the muscle loosens its chronic grip. You can forage for relief to give your "mind muscle" a break, then forage for insight to figure out how to let it relax. Once you're on a foraging path, you can consider what purpose foraging can serve. Let's look at this in action.

"It figures," Brendan thought mockingly, "this guy, a famous hypnotist, feels he has carte blanche to interpret other people's behavior. . . . What a clown." Brendan was showing tonight's performer, a famous hypnotist, around the venue that he manages. He was fiddling with the door keys when the hypnotist casually remarked, "You seem to have an issue with keys and doors; were you often locked in your room as a child?" It didn't help that Brendan also fumbled with unlocking his car door when he picked the hypnotist up at the hotel earlier that morning. But assuming childhood trauma . . . really?

Brendan felt triggered by the hypnotist's offhand remark; he decided to go for a brief walk and take in the local park by the theater. After a few minutes of letting the sights and sounds of nature occupy his mind, the incident seemed less important, and he went on with his day.

A few weeks later, Brendan was paying for parking on a rainy fall morning when he was reminded of the hypnotist's observation. Just like his keys, he couldn't get the parking pay station to operate. Standing in the cold rain in front of the machine, he repeatedly entered his license plate info on a touch screen, but it seemed like the information wasn't being accepted. He noticed himself becoming frustrated, just like when he couldn't get the theater and car doors open on his first try. "The machine must be broken. I'd rather risk getting a ticket than deal with this bullshit."

Brendan stepped away for a moment to take a breath and consider whether it really felt right just to walk away. When he turned back to the machine, he noticed that his license plate information was displayed, but in a place he hadn't been looking—at the very *top* of the rain-covered screen. "Heh, it was there all along, but I kind of blanked out and stopped looking." He then remembered the same feeling of "blanking out" in the past when he got stuck on crossword puzzles or the time his nephew asked him to solve a Rubik's Cube. Usually, this feeling came right before feeling frustrated and giving up.

Allowing this realization to sink in, Brendan searched his body for what "blanking out" felt like. He felt tightness in his forehead, squinting, and his heart beating a bit faster, kind of pounding in his temples. "I wonder what would happen if the next time that I feel locked out or that I can't see a solution, I check for this 'blanking out' sensation. It might help me to take a step back and see things that I am missing but were there all along."

In this example, it would have been great if Brendan could have simply taken a breath and gracefully unlocked doors, or felt the wind on his cheek and seen the parking machine as some fun puzzle to solve, an unexpected Wordle or sudoku gifted to him. But it didn't play out that way. By the time Brendan knew what was going on, his nervous system had already been triggered, stress had taken hold, and the default mode network was spitting out defensive solutions tinged with anger and resentment. To simply say "I'm foraging and therefore happy" would have been inauthentic.

On the other hand, foraging *did* help. It gave Brendan a place to go with his mind that wasn't just banging his head against the wall of "it ought not to be so." Instead, he was able to let go of active inference and value perceptual inference for a few moments. He was able to connect with a live, pulsing sensory experience that gave him something else to focus on; there's a relief in putting down your frustration for a moment and slipping into an exploratory mode. And when he returned, he was able to see more than just the failure of his default model, which is to rush to get things done. By taking in more of the situation, he could slow down and get the key in the lock, notice a cue from the parking-machine screen in an unexpected place. For Brendan, sense foraging offered a constructive disruption to the status quo.

What's more, repeated engagement with these problems planted the seeds of insight. When dealing with the parking meter, Brendan started to grow familiar with the feeling of frustration, the "blanking out" sensation. He was able to take the discovery and be on the lookout for blanking out in the future. This cue will help him recognize the need to slow down sooner going forward; he has not yet escaped the habit of rushing, but he is no longer as fated to continue rushing his way into frustration and anger.

PICK A PEAK

Having a target is critical. Brendan was grappling with rushing into frustration. Shanice wanted to be able to run her business successfully without running herself into the ground. Neither of them knew how to do it; all they yearned for in the moment was escaping that stuck feeling. But each of them had a goal, as did Newton trying to figure out how the world works, and Archimedes discerning solid gold from dross. They simply didn't know how to get there. On the other hand, they understood that something was missing and determined to try to find it.

While it's important to have clarity about what's missing in your life, this doesn't mean "oh, if I only had a winning lottery ticket" or "if only I could get with that hot new marketing manager." What we mean is that, in your heart of hearts, when you aren't happy, there's a *feeling* that's missing (Updike's "That's it!"), and its absence may be scary to admit. Maybe you don't feel calm enough, respected enough, smart enough, successful enough, or loved enough. Maybe you wish you could feel joy or excitement or even some sort of romantic yearning that the poets write about. If you were to take a quiet moment, you might sense something missing that gives rise to your sadness, irritation, anger, or distraction. And it would be scary to admit that this yearning is running you—because what happens if you can't ever achieve that feeling of completeness and sufficiency?

But worrying about not finding what you desire is a one-way trip back into the House of Habit, pulling down the shades, and locking the door in case failure comes knocking. The fisherman doesn't avoid his boat for fear of not catching fish; the accountant doesn't avoid spreadsheets for fear of making an error. When you begin to sense forage, you've got to be brave enough to stand for something, even though it opens you up to the possibility of failure.

As we've said before: running toward beats running from. This truth is also at the cutting edge for helping people recover from substance use disorders. The "secret" is not to figure out why one became an addict, or to break the addictive circuit by pumping up some inhibitory muscle. Hoping your willpower will be strong enough to allow you to resist your drug of choice doesn't work. In the long run, attraction to something beats resistance against it every time.

While in the short term, just saying no is fine, it doesn't change the underlying brain circuitry, so when stress comes—and it always does—stress will do what it does, shutting down sensation and locking you back into habit, exactly where you were trying to escape from in the first place. You need to rebuild the reward circuits that have been hijacked by drug use, and for that, sense foraging is key.

Scans of the brain's reward circuity in cocaine users showed reduced responses to the promise of nondrug rewards such as money.[14] Even though money can buy drugs, addicts' brains have been so focused, their House of Habit so carefully curated to serve the addiction, that only drugs and drug paraphernalia provoke the slightest activation.

Fortunately, even addicts can form new habits, ones that support rather than erode well-being—but it takes time.[15] One month into abstinence, a meth addict's reward regions remain almost silent during tests of coordination and memory compared to someone who has never used meth. After a year of abstinence, however, activation in these regions begins to recover. On PET scans, one can see new reward responses forming, like shoots of grass poking through the earth after a long winter. Such changes are the neural signpost of a person learning to find meaning in everyday comforts once more. But a year! That's a lot of time to spend redecorating your House of Habit, just saying no to your only source of pleasure. What do you do while waiting for your brain to recover?

This is where the real work is. Once you're equipped with knowledge of sense foraging, choosing something to forage for—your core values—is the final step that will enable you to begin to change what motivates your life. To explore this process, let's try the following exercise:

Exercise: Write Your Own Obituary (with a Nod to Alfred Nobel)

Imagine that you're looking down from above, shortly after your death, curious about how you're being remembered. This can be years down the road from now—you've had time to accomplish things, to have grown as a person, to have improved the nature of your relationships. Let's assume you've done what you wanted with your life, thanks to a combination of your own hard work and the kinds of insights we've provided in this book. So, what does your obituary say about you?

Long before there was GPS, even before there was the sextant or the compass, ancient mariners practiced celestial navigation, relying on "heavenly bodies" to guide them when sailing beyond sight of land. Often, they would sail toward a particular star, called a lodestar.

The obituary you imagine for yourself—and what you want it to say—can be your lodestar for navigating the sea of life. Does it describe you as having been devoted to your family, or does it say you made a lot of money? Does it say you were spiritually centered? The life of the party? Were you generous, a greedy bastard, driven to achieve, committed to social justice? Did you devote your time on earth to creating great art, or did you watch a lot of Netflix?

Sailing on the open ocean without being sure you're on the right course is incredibly stressful. But once you set your sights on the constellation of values and commitments you want your life to be about, most of the stress fades away. If you stay the course, your chosen destination as your lodestar, the journey makes sense, and the struggles become less of a burden.

So now, please take ten minutes and write your obituary. You don't have to show it to anyone, and you don't have to polish it to make it sound perfect. Just write it, and then read it over and examine the values you featured when describing your life as you want to live it. Can you raise those values up to be your lodestar, guiding you across uncharted, sometime turbulent seas? Remember, a value can have many different manifestations. You don't have to give away all you have to embody the value of generosity. You don't have to go work in a refugee camp to embody the value of service.

A NEW KIND OF MASTERY

Like Alfred Nobel, you've just had the opportunity to see your life summed up while you're still living it. And like Nobel, you can use the experience to refine and solidify your values and your commitment to them. Sense foraging can be the first step, and there is no right way or wrong way when you set out to explore. It's not about mastery; it's about discovery.

Ironically, your first step into sense foraging will very likely make things worse. If you're like Shanice, your habits have served you well—up to the point at which you ceased flourishing and started languishing. You got stuck, but along the way, you got your degree, or your middle-management job, or found what seemed like a satisfying relationship. In other words, you've reached what mathematicians would call a "local maximum," a place where things seem about as good as they can get because your approach to further improvement is limited.

We can see this theme arise repeatedly in popular books and movies. In *Dirty Dancing* the "Baby" who no one puts in a corner risks the comfortable middle-class life her parents envision for her in the pursuit of love with a blue-collar worker who believes in her. Jack Nicholson's character in *As Good as It Gets* and the old man from *Up* are both older adults set in their ways, who have purposely fortified themselves against outside intrusions, only to realize they are missing out on life itself. In each of these examples, the characters begin the story caught in local maxima, places where things are safe, controllable, but ultimately unsatisfying.

In an unpredictable world with vague threats lurking in the shadows, it seems reasonable to circle the wagons and try to protect what you've got. But this conservative impulse means accepting a fixed level of happiness, feeling more anxiety about change than excitement about finding opportunities to grow. We get stuck in a local maximum because any change, even one that will appear totally necessary in hindsight, requires a true loss of stability before you make it. Destabilizing a system that kind of works to forage for something better means sliding a few steps back down Maslow's pyramid and being forced to work your way back up a new learning curve.

You might be getting passable results with your golf grip, your backhand, or the way you hold the neck of your guitar, but your instructor tells you that you'll never really improve unless you switch to the preferred way. If you bite the bullet and make the change, your performance will likely decline in the short run as you try to master what seems an awkward modification. Within a few days or weeks, though, you realize that you're now on the path to higher ground. Anytime you're on the "optimization peak" of a local maximum (no matter how paltry), you can't get to a higher peak without first going down into the surrounding valley.

If you understand this idea of local maxima, you can prepare yourself for feeling a bit confused and uncertain as you begin to forage for what's missing in your life. You've worn out your old rules of thumb, so you'll be *less efficient* in certain familiar routines as you start to consider new possibilities. But being inefficient is another way of saying that you're shopping around and trying out new things. And unless you really think what you've got is "as good as it gets," experimentation and exploration are the only ways to discover what else the world has to offer.

DO SOMETHING WEIRD

When you're stuck in a local maximum and want to get unstuck, there's a big difference between jumping off the mountain and backtracking to find a new path to higher ground. It's the difference between acting on impulse and taking the *sensible* approach to changing your life.

Consider the example of Arunima.

Arunima loved walking in the early evening. Not only did it help her sleep more soundly, but she was only a few minutes away from a trail through a wooded park. One evening, she felt a cool breeze on her face as she wound her way home after another refreshing outing. Walking up her street, she noticed a beer can and a take-out cup that had been tossed on the curb. "Wow, I guess some people just don't care about anyone but themselves," she thought.

Arunima had always felt that, if we all played a role in looking after our communities, things could change for the better. Then it occurred to her, "Why not just pick up this junk, carry it home, and put it in the recycling bin?" This was quickly countered by "because people will think I'm homeless or deranged." Her DMN resisted violating social norms, and the normative habit was to go "tut-tut," perhaps, but to simply keep on walking. But then it occurred to her, "If this kind of trashy behavior really bothers me—runs counter to what I value—here's a chance to do something. I can come up with lots of reasons for not doing it, but something just doesn't feel right about walking away."

Aware of both her conviction and her apprehension, Arunima surprised herself by bending down and picking up the can and the cup, then carrying them to her house and recycling them. "It may not be much in the scheme of things, but I was able to make a small difference today," she thought afterward.

HOW TURNING IN TURNS OUT

When you let sense foraging lead you away from habit and onto an unconventional path, you might find that you aren't alone on your journey toward personal growth.

A few days after her walk, Arunima casually told her friend Zahra how weird it felt to bend down and pick up other people's garbage. She had no regrets, however, because just walking past it would have felt worse. Zahra, to her surprise, said, "Cool, maybe I'll come with you next time." What Arunima didn't know is that behavior you fear might look crazy to others might be on other people's minds as well, everyone waiting for that one habit breaker to create a space for connection and engagement. This is precisely how the social movement #trashtag—whose motto is "picking up the planet"—got its start.

In 2019, Byron Roman used social media to challenge teens to pick up litter, acting locally to clean up the planet.[16] He got their attention by posting before and after photos of Drici Tani Younes cleaning up a local road near his home in Algeria.[17] To his amazement, the post went viral, with 325,000 shares on Facebook and more than 25,000 posts on Instagram. It wasn't long before people were posting similar photos of their cleanup efforts, and the #trashtag challenge was embraced worldwide. Today, a vibrant community has developed around a simple decision to break with habit.[18]

But social movements don't begin with hundreds of thousands of participants. They begin with one person who is willing to engage in sense foraging and accept what they find. The reward is not the social media following; we suggest instead that the following reflects how available foraging insights are to others around the world once they are given a nudge in the right direction.

All of us assume that others know what makes us tick, but the reality is that we can only observe external action. Who knows how many people of color thought about violating Montgomery's segregation laws before Rosa Parks took her seat in the front of the bus. Someone must have the courage to go first in defying the habitual. Then they can create a visible rallying point for change.

The goal of living a sensible life is not necessarily to lead a social movement, and it's unlikely that your choices will go viral. But we believe that anything you do in response to sincere foraging can act as a beacon for others. Your action can be another mariner's lodestar and change the communities to which you belong. As you forage for what you value, you will discover not only ideas but also people who are foraging for the same thing. And meeting those people is another wellspring that can sustain you in the long run.

TURNING BACK IN

Arunima never became famous for her foraging-turned-cleanup efforts. But that was fine; the unexpected satisfaction of taking a risk and having it turn out okay stayed with Arunima for the rest of the day. By the end of the week, she found herself wondering where else foraging could lead. Within two weeks of exploring her feelings at work and at home, Arunima was playing with the idea of what breaking out of the habitual, rule-following mold would look like in the context of her work.

Being a freelance editor allowed her to set her own hours, work from home, and choose projects that she found intellectually stimulating. In the last few years, though, her field had become increasingly automated, outsourced, and impersonal. Managers she barely knew were hounding her to complete assignments more quickly, lower her hourly rate, or take on extra projects that would have her working twenty-hour days.

In her head, Arunima always thought her options were either just "take it" or line up another job. But thinking it through this time, she realized that as a contractor working on a project-by-project basis, moving to another company would simply relocate, and re-create, the same old grind. What really needed to change was her!

According to the conventional wisdom, quitting her job would just make things worse by depriving her of income, which would just add to her stress. But when Arunima foraged for inspiration, she recognized the feeling that came as "exhaustion." Rather than helping her power through, Arunima's insight was that conventional wisdom didn't really apply here. She needed to follow unconventional wisdom to escape the habits she'd fallen into, to reclaim the version of Arunima who had ambitions, dreams, and spark. In her current, worn-down mental rut, she couldn't even muster the motivation to look for another job. She decided to follow a middle path, sticking a toe into uncertainty by taking a leave of absence rather than opting out entirely. She resolved to use that breathing room to assess the bigger picture without the stress of her current situation. In doing so, she still had to leave the "local maximum" of a steady income but was willing to tolerate this stress for a few months in return for the possibility of changing her fate.

Going into her leave, Arunima had no idea how things would play out, and she realized that this was the first time she'd had this kind of open-ended view of the future in a long time. It was exhilarating.

Gradually, perceptual inference snuck into her House of Habit, and she began to remember why she'd gone into editing in the first place. She'd been inspired to curate great works of literature, but she'd fallen into a rut in which her greatest contribution was reminding academics to refrain from splitting infinitives. Yet even though the flame had been diminished, it hadn't been extinguished—she still yearned to work with great writers. She knew she'd probably never be working at *The New Yorker*, but she needed direction in her life, so she might as well steer toward something she really cared about.

While it took some time for Arunima to find her lodestar, in the following weeks she regained the motivation that her old routine had beaten out of her. She began networking to find projects far more creative than the technical editing she'd been doing. She explored editing film and television scripts, she volunteered to judge a poetry contest, and she began editing copy for a large auction house that represented prestigious works of art. Once she understood that her goal was to contribute to beauty in the world, the specific avenues she took to get there seemed less important. What mattered was that she was on her way.

MAKING SENSE OF IT ALL

At the beginning of this book, we said we weren't going to give you a money-back guaranteed, predictable program to follow. Instead, we said we would challenge you to do something different, and on your own terms.

Give yourself permission to be stuck and notice it. Try sense foraging just to see what it's like. Toggle between conceptual and sensory modes and see what unfolds. Write your obit and note what was important in the end. Take one weird action.

Years from now, at the very least, your obituary will read: you cared about something and had the courage to seek it.

After all, isn't that the sensible thing to do?

Exercise: Putting It All Together

Let's finish our journey together by reviewing the map that we have constructed. You will notice that the destination is intentionally left blank; to set off on your own journey means embracing the unknown. But there are many markers to help you on this journey. Let's review:

1. **Get to know your senses.** Learn how to make the "move" of attending to sensation and letting the influx of sensory information, rather than some later inference or change, be your criterion for success.
2. **Find ways to connect with your senses even in the business of everyday life.** Practice tuning into sensation not just when you feel amazing and are in a quiet and tranquil space, but amid the busyness of everyday life as well—feel your feet on a crowded street, the wind when you first step outside, the sight and sound of people's faces and voices even though you have tons to do. Most of all, get used to checking in with the sensations arising inside

your body, the building blocks of feeling, which in turn become appraised as full-blown emotions.

3. **Notice that some sensations form familiar patterns that are predictable.** There is one feeling we really want you to get to know: the feeling of being stuck. We know, it is probably one of the least pleasant feelings you can become familiar with, yet many of us are spending a great deal of our day in this place. Getting to recognize being stuck, and feeling good about noticing, is the first step to getting unstuck.
4. **When you feel stuck, learn to forage for change.** They say the only thing that never changes is the fact that things change, but we lose track of this knowledge when we become stressed. When you are feeling stuck, notice that almost all your sensory information shows a world that is dynamic, alive, and changing. If you can attend to other sensations in your own body beyond the constellation of sensations that give rise to the feeling of stuckness, you can see that you are not truly locked in place—that's just a description of an overly familiar unpleasant sensation. Get used to responding to feelings of stress and being stuck by tuning into the experience of change all around you, and even inside you.
5. **Notice that dynamic sensations contain a universe of feeling tones.** Once you get used to noticing the snap, crackle, and pop of changing sensations, you can notice them under the dominant feeling of stress on loop. Yet there are all kinds of other feelings, too—of wanting more of a pleasant experience, less of an unpleasant one, or even moments of indifference—that ask for no action at all. Can

you learn to notice these tones, to recognize that there is never only one feeling in play? With so many avenues for feeling that we could choose to bring to the surface?

6. **Make a game of sense foraging in a new location.** Get used to moving into new locations and foraging for new, commonly ignored sensations. If you are at the grocery store, look at foods you normally ignore. Snoop a bit at what other people are buying. See if you can figure out the mood of the cashier before you get to the front of the line. See if you can play games of noticing the oft-ignored aspects of your experience wherever you go.
7. **Get intentional about sense foraging.** As you get to know deep-set patterns and habits, you will need to get more intentional and plan sense-foraging activities. Notice for yourself how deeply caught in a mood or train of thought you are, and use that knowledge as motivation to choose a safe foraging activity. Set a time and place and, most important, go out and do it—even if it feels strange and awkward; those are often signs you are doing something new and you are on the right track!
8. **Find your access points.** Don't forage the same way someone else does just because it works for them. Think about the different access points we've suggested, or think of some activities that you already enjoy that could serve as sense-foraging exercises with the right intention and attention to sensation. Just like a person learning a few tried-and-true cooking recipes, a resilient person will have a few different access points for sense foraging, some that provide a sense of relief, some that help build insight, and some that have fringe benefits like group activities that

strengthen relationships with others, or fitness activities that are good for your health.

9. **Remember to toggle back to your problems.** Don't fall into the trap of only using sense foraging for relief while real problems build around you. Learn to forage for relief when you are overwhelmed, but also make time each week to forage for insight—some proactive foraging is needed to solve problems, to have that eureka moment, rather than simply pumping the brakes on the road to burnout with brief reprieves into relief. Come back to the conceptual and notice if other options are available after sense foraging, and if you are still stuck, then forage again. Get used to toggling between these modes to keep your eye on the important things in life while you also work to keep yourself relaxed, open, and flexible through sense foraging.
10. **Think about the big picture.** As you get more comfortable with sense foraging and toggling, you will feel like you sometimes have a greater capacity to think about the big picture rather than just survival. Maybe you clean up trash because it disgusts you personally, and maybe you pick up trash because you want to live in a cleaner world. This is another form of toggling, but when you have the wherewithal, challenge yourself and ask: Can I play a bigger game than I am already playing? Can I tackle bigger problems? Sometimes the answer will be no, but you may find to your surprise that you are up for more than you would normally expect for yourself.

Acknowledgments

ACADEMIA IS FULL OF SILOS. ONE COULD SPEND AN ENTIRE career improving the treatment of depression or probing the neuroscience of body awareness, without these two streams of research ever intersecting. We were exceptionally lucky to be drawn together into the mystery of sensation, which surprisingly connected our two silos. Who would have thought that two decades ago, as Norm first appeared in Zindel's office at the Centre for Addiction and Mental Health, that we would become regular collaborators and ultimately friends and coauthors? Our acknowledgments must begin with gratitude that our seemingly disparate worldviews had the opportunity to coalesce into a message greater than the sum of its parts.

The bridge builder in this case was Professor Adam K. Anderson, who attended a lecture Zindel gave at the University of Toronto and shared his interest in using fMRI to investigate mindfulness meditation to prevent relapse in depression. As Adam's graduate student, Norm regularly attended the meetings to develop the first imaging studies discussed in this book, and we are grateful to Adam for creating this connection. There aren't many prominent cognitive neuroscientists who have the breadth of vision to tackle something as ephemeral as meditation, but Adam was up for the challenge.

It took a village to raise us—a wave of courageous contemplative teachers and scholars who showed that science and contemplative practice could be fruitfully combined. Teachers like Jon Kabat-Zinn, Sharon Salzberg, Joseph Goldstein, Satya Narayana Goenka, Ajahn Amaro, Jack Kornfield, Matthieu Ricard, Trudy Goodman, and His Holiness the Dalai Lama himself all contributed meaningfully to our own contemplative practices, but also to more formal dialogues with modern scientists and clinicians, often orchestrated through the Mind and Life Institute. Similarly, scientists like Ritchie Davidson, Cliff Saron, Sara Lazar, Catherine Kerr, Antoine Lutz, and Wolf Mehling powerfully demonstrated how to apply rigorous methods to develop a new contemplative science. In sensory neuroscience, Bud Craig, Hugo Critchley, Helen Mayberg, and Yvette Sheline breathed new life into the neuroscience of body awareness. In clinical science, Mark Williams and John Teasdale, Sona Dimidjian, Cynthia Price, and Marsha Linehan conceived and validated secular, accessible means for bringing contemplative practice into the sphere of clinical care.

We are blessed with wonderful colleagues at the University of Toronto: fellow faculty, postdocs, graduate students, research assistants, and trainees in our labs, who keep us engaged, challenged, and growing throughout the years. You have all been effective sounding boards and engaged interlocutors and have reeled our ideas in when they became too speculative or extreme, but also encouraged us toward boldness when warranted. Beyond our institution, there are so many other incredibly brilliant and generous minds that have influenced us in both direct interactions and in benefiting from their scholarship. In this as in all things, we are all interconnected.

The research that has challenged and updated our own models of the world would not have been possible without generous support from so many sources. We are grateful for funding from the Natural Sciences and Engineering Research Council (NSERC), the Social Sciences and Humanities Research Council (SSHRC), the Canadian Institutes of Health Research (CIHR), the National Institute of Mental Health (NIMH), the Mind and Life Institute, the MacArthur Foundation, the Ontario Mental Health Foundation, and the 1440 Foundation.

Our first agent in this process was Linda Loewenthal, who provided critical initial feedback and support for the book concept. When life conspired against our collaboration, Linda graciously connected us to agent extraordinaire Carolyn Savarese, who helped us focus on neuroscience, sense foraging, and the potential for connecting a mass audience to a new approach to well-being. Carolyn has supported us in all the right ways: she's been tough in pushing us to do better while nurturing our creative growth. Our good fortune continued in connecting us with William Patrick, an accomplished writer and editor, who taught us to trim away academic fat to reveal the lean sinew of the narrative you see before you today. In our eyes, getting to work with Carolyn and Bill felt like getting selected for an all-star team—in the uncertain process of conceiving this book, their belief in this project fueled our own.

We are incredibly grateful to have partnered with Tracy Behar and Little, Brown Spark. She saw value in our vision from the outset and helped us to zero in on sense foraging as the central concept. From our first meeting, we became quick admirers of her wide-ranging experience and judgment. Equally comfortable with conceptual debates and with the pragmatics of the production process, Tracy has provided steadfast guidance every step of the way.

We would never have had the space, the motivation, or the good humor to write this book without the unflagging support of our families and friends. For Norm, his partner, Kate, and his children, Quentin and Phoenix, and for Zindel, his wife, Lisa, and his children, Ariel, Shira, and Solomon, are the foundation from which everything is possible.

Finally, for all you new foragers out there: we hope your future is deliciously surprising.

Notes

CHAPTER 1

1. Eva de Mol et al., “What Makes Entrepreneurs Burn Out?,” *Harvard Business Review*, April 4, 2018, https://hbr.org/2018/04/what-makes-entrepreneurs-burn-out.

2. Ashley Abramson, “Burnout and Stress Are Everywhere,” American Psychological Association, January 1, 2022, https://www.apa.org/monitor/2022/01/special-burnout-stress.

3. Kristy Threlkeld, “Employee Burnout Report: COVID-19’s Impact and 3 Strategies to Curb It,” Indeed for Employers, March 11, 2021, https://uk.indeed.com/lead/preventing-employee-burnout-report.

4. Melanie Anzidei, “Naomi Osaka’s Decision to Step Away from Tennis Shines Light on Athlete Burnout,” NorthJersey.com, September 11, 2021, https://www.northjersey.com/story/sports/tennis/2021/09/11/naomi-osaka-us-open-2021-loss-leylah-fernandez/8278595002/.

5. Jennifer Berg, “Ipsos U.S. Mental Health 2021 Report,” Ipsos, May 21, 2021, https://www.ipsos.com/en-us/news-polls/ipsos-us-mental-health-2021-report.

6. Georgia Wells, Jeff Horwitz, and Deepa Seetharaman, “Facebook Knows Instagram Is Toxic for Teen Girls, Company Documents Show,” *Wall Street Journal*, September 14, 2021, https://www.wsj.com/articles/facebook-knows-instagram-is-toxic-for-teen-girls-company-documents-show-11631620739.

7. P. Fossati et al., "In Search of the Emotional Self: An fMRI Study Using Positive and Negative Emotional Words," *American Journal of Psychiatry* 160, no. 11 (2003): 1938–1945.

8. Damien Gayle, "Facebook Aware of Instagram's Harmful Effect on Teenage Girls, Leak Reveals," *The Guardian*, September 14, 2021, https://www.theguardian.com/technology/2021/sep/14/facebook-aware-instagram-harmful-effect-teenage-girls-leak-reveals.

9. F. I. Craik and R. S. Lockhart, "Levels of Processing: A Framework for Memory Research," *Journal of Verbal Learning and Verbal Behavior* 11, no. 6 (1972): 671–684.

10. N. A. Farb et al., "Attending to the Present: Mindfulness Meditation Reveals Distinct Neural Modes of Self-Reference," *Social Cognitive and Affective Neuroscience* 2, no. 4 (2007): 313–322.

11. C. G. Davey et al., "Mapping the Self in the Brain's Default Mode Network," *NeuroImage* 132 (2016): 390–397.

CHAPTER 2

1. M. E. Raichle et al., "A Default Mode of Brain Function," *Proceedings of the National Academy of Sciences* 98, no. 2 (2001): 676–682.

2. J. Smallwood and J. W. Schooler, "The Science of Mind Wandering: Empirically Navigating the Stream of Consciousness," *Annual Review of Psychology* 66 (2015): 487–518.

3. T. J. La Vaque, "The History of EEG Hans Berger: Psychophysiologist; A Historical Vignette," *Journal of Neurotherapy* 3, no. 2 (1999): 1–9.

4. Blaise Pascal, "Pensées," University of Fribourg (in French). Archived (PDF) from the original on January 24, 2004.

5. L. Sokoloff et al., "The Effect of Mental Arithmetic on Cerebral Circulation and Metabolism," *Journal of Clinical Investigation* 34, no. 7 (1955): 1101–1108.

6. D. H. Ingvar, "Memory of the Future: An Essay on the Temporal Organization of Conscious Awareness," *Human Neurobiology* 4, (1985): 127–136.

7. Z. V. Segal et al., "Cognitive Reactivity to Sad Mood Provocation and the Prediction of Depressive Relapse," *Archives of General Psychiatry* 63, no. 7 (2006): 749–755.

8. Y. I. Sheline et al., "The Default Mode Network and Self-Referential Processes in Depression," *Proceedings of the National Academy of Sciences* 106, no. 6 (2009): 1942–1947.

CHAPTER 3

1. Nicola Palomero-Gallagher and Katrin Amunts, "A Short Review on Emotion Processing: A Lateralized Network of Neuronal Networks," *Brain Structure and Function* 227 (2022): 673–684, https://link.springer.com/article/10.1007/s00429-021-02331-7.

2. N. A. Farb et al., "Minding One's Emotions: Mindfulness Training Alters the Neural Expression of Sadness," *Emotion* 10, no. 1, (2010): 25.

3. G. Sheppes et al., "Emotion Regulation Choice: A Conceptual Framework and Supporting Evidence," *Journal of Experimental Psychology: General* 143, no. 1 (2014): 163.

CHAPTER 4

1. N. A. Farb, Z. V. Segal, and A. K. Anderson, "Attentional Modulation of Primary Interoceptive and Exteroceptive Cortices," *Cerebral Cortex* 23, no. 1 (2013): 114–126.

2. N. A. Farb, Z. V. Segal, and A. K. Anderson, "Mindfulness Meditation Training Alters Cortical Representations of Interoceptive Attention," *Social Cognitive and Affective Neuroscience* 8, no. 1 (2013): 15–26.

CHAPTER 5

1. L. F. Barrett, "The Theory of Constructed Emotion: An Active Inference Account of Interoception and Categorization," *Social Cognitive Affective Neuroscience* 12, no. 1 (2017): 1–23, https://www.ncbi.nlm.nih.gov/pmc/articles/PMC5390700/.

2. M. A. Goodale and A. D. Milner, "Separate Visual Pathways for Perception and Action," *Trends in Neurosciences* 15, no. 1 (1992): 20–25, https://doi.org/10.1016/0166-2236(92)90344-8.

3. Link R. Swanson, "The Predictive Processing Paradigm Has Roots in Kant," *Frontiers in Systems Neuroscience* 10 (2016), https://doi.org/10.3389/fnsys.2016.00079.

4. Juliet Macur, "Scott Hamilton Was Demoted as an Olympic Broadcaster. Don't Feel Sorry for Him," *New York Times*, February 18, 2018.

5. Carl R. Rogers, *On Becoming a Person: A Therapist's View of Psychotherapy* (New York: Houghton Mifflin, 1961).

6. N. Farb et al., "Interoception, Contemplative Practice, and Health," in "Interoception, Contemplative Practice, and Health," special issue, *Frontiers in Neuroscience* 6 (2015).

CHAPTER 6

1. Z. V. Segal et al., "Practice of Therapy Acquired Regulatory Skills and Depressive Relapse/Recurrence Prophylaxis following Cognitive Therapy or Mindfulness Based Cognitive Therapy," *Journal of Consulting and Clinical Psychology* 87, no. 2 (2019): 161.

2. Christine A. Padesky, "Socratic Questioning: Changing Minds or Guiding Discovery?" (keynote address presented at the European Congress of Behaviour and Cognitive Therapies, London, 1993), https://padesky.com/newpad/wp-content/uploads/2012/11/socquest.pdf.

3. N. A. Farb et al., "Static and Treatment-Responsive Brain Biomarkers of Depression Relapse Vulnerability following Prophylactic Psychotherapy: Evidence from a Randomized Control Trial," *NeuroImage: Clinical* 34 (2022): 102969.

CHAPTER 7

1. A. Drake et al., "Daily Stressor-Related Negative Mood and Its Associations with Flourishing and Daily Curiosity," *Journal of Happiness Studies* 23, no. 2 (2022): 423–438.

2. C. J. Burnett et al., "Need-Based Prioritization of Behavior," *Elife* 8 (2019): e44527.

3. Reporter's Notebook, "Firewalking: Mind Over Matter or a Tool for Personal Growth, or Both?," ABC News, November 5, 2009, https://abcnews.go.com/Health/firewalking-mind-matter-tool-personal-growth/story?id=9011713.

4. Andrew Weil, foreword to *Extreme Spirituality: Radical Approaches to Awakening*, by Tolly Burkan (Chicago: Council Oak Books, 2004).

5. Helen S. Mayberg et al., "Deep Brain Stimulation for Treatment-Resistant Depression," *Neuron* 45, no. 5 (2005): 651–660, https://www.sciencedirect.com/science/article/pii/S089662730500156X.

CHAPTER 8

1. J. F. Kelly et al., "How Many Recovery Attempts Does It Take to Successfully Resolve an Alcohol or Drug Problem? Estimates and Correlates from a National Study of Recovering US Adults," *Alcoholism: Clinical and Experimental Research* 43, no. 7 (2019): 1533–1544, https://doi.org/10.1111/acer.14067.

2. O. Sacks, *Everything in Its Place: First Loves and Last Tales* (London: Picador, 2019).

3. Kathryn E. Schertz and Marc G. Berman, "Understanding Nature and Its Cognitive Benefits," *Current Directions in Psychological Science* 28, no. 5 (2019): 496–502, https://journals.sagepub.com/doi/10.1177/0963721419854100.

4. Mathew P. White, "Spending at Least 120 Minutes a Week in Nature Is Associated with Good Health and Wellbeing," *Scientific Reports* 9, no. 7730 (2019), https://www.nature.com/articles/s41598-019-44097-3.

5. C. L. Anderson, M. Monroy, and D. Keltner, "Awe in Nature Heals: Evidence from Military Veterans, At-Risk Youth, and College Students," *Emotion* 18, no. 8 (2018): 1195–1202, https://psycnet.apa.org/record/2018-27538-001.

6. Robert Moor, "When Thoreau Went Nuts on Maine's Mt. Katahdin," *Sierra*, August 19, 2016, https://www.sierraclub.org/sierra/2016-4-july-august/green-life/when-thoreau-went-nuts-maine-s-mt-katahdin.

7. Zhi Cao et al., "Associations of Sedentary Time and Physical Activity with Adverse Health Conditions: Outcome-Wide Analyses Using Isotemporal Substitution Model," *eClinicalMedicine* 48 (2022): 101424, https://www.sciencedirect.com/science/article/pii/S2589537022001547.

8. Juan Gregorio Fernández-Bustos et al., "Effect of Physical Activity on Self-Concept: Theoretical Model on the Mediation of Body Image and Physical Self-Concept in Adolescents," *Frontiers in Psychology* 10 (2019), https://doi.org/10.3389/fpsyg.2019.01537.

9. Qingguo Ding et al., "Sports Augmented Cognitive Benefits: An fMRI Study of Executive Function with Go/NoGo Task," in "Stress-Induced Glial Changes in Neurological Disorders 2021," special issue, *Neural Plasticity* (2021), https://www.hindawi.com/journals/np/2021/7476717/; E. I. de Bruin, J. E. van der Zwan, and S. M. Bögels, "A RCT Comparing Daily Mindfulness Meditations, Biofeedback Exercises, and Daily Physical Exercise on Attention Control, Executive Functioning, Mindful Awareness, Self-Compassion, and Worrying in Stressed Young Adults,"

Mindfulness 7 (2016): 1182–1192, https://link.springer.com/article/10.1007/s12671-016-0561-5.

10. Costas I. Karageorghis, Symeon P. Vlachopoulos, and Peter C. Terry, "Latent Variable Modelling of the Relationship between Flow and Exercise-Induced Feelings: An Intuitive Appraisal Perspective," *European Physical Education Review* 6, no. 3 (2000): 230–248, https://journals.sagepub.com/doi/abs/10.1177/1356336X000063002.

11. Ernst Pöppel et al., "Sensory Processing of Art as a Unique Window into Cognitive Mechanisms: Evidence from Behavioral Experiments and fMRI Studies," *Procedia—Social and Behavioral Sciences* 86 (2013): 10–17, https://www.sciencedirect.com/science/article/pii/S1877042813026487.

12. Chun-Chu Chen and James F. Petrick, "Health and Wellness Benefits of Travel Experiences: A Literature Review," *Journal of Travel Research* 52, no. 6 (2013): 709–719, https://journals.sagepub.com/doi/full/10.1177/0047287513496477.

13. IpKin Anthony Wong, Zhiwei (CJ) Lin, and IokTeng Esther Kou, "Restoring Hope and Optimism through Staycation Programs: An Application of Psychological Capital Theory," *Journal of Sustainable Tourism* 31, no. 1 (2023): 91–110, https://www.tandfonline.com/doi/full/10.1080/09669582.2021.1970172.

14. Catherine I. Andreu et al., "Enhanced Response Inhibition and Reduced Midfrontal Theta Activity in Experienced Vipassana Meditators," *Scientific Reports* 9, no. 1 (2019): 13215, https://doi.org/10.1038/s41598-019-49714-9.

15. *Online Etymology Dictionary*, s.v. "patience," https://www.etymonline.com/search?q=patience.

16. G. Petri et al., "Homological Scaffolds of Brain Functional Networks," *Journal of the Royal Society Interface* 11, no. 101 (2014): 20140873.

17. T. Anderson et al., "Microdosing Psychedelics: Personality, Mental Health, and Creativity Differences in Microdosers," *Psychopharmacology* 236 (2019): 731–740.

18. Hugo Bottemanne et al., "Evaluation of Early Ketamine Effects on Belief-Updating Biases in Patients with Treatment-Resistant Depression," *JAMA Psychiatry* 79, no. 11 (2022): 1124–1132, https://doi.org/10.1001/jamapsychiatry.2022.2996.

19. T. Anderson et al., "Psychedelic Microdosing Benefits and Challenges: An Empirical Codebook," *Harm Reduction Journal* 16, no. 1 (2019): 1–10.

20. N. Gukasyan and S. M. Nayak, "Psychedelics, Placebo Effects, and Set and Setting: Insights from Common Factors Theory of Psychotherapy," *Transcultural Psychiatry* (2021): 1363461520983684.

CHAPTER 9

1. D. Frye, P. D. Zelazo, and T. Palfai, "Theory of Mind and Rule-Based Reasoning," *Cognitive Development* 10, no. 4 (1995): 483–527.

2. A. Tymula et al., "Adolescents' Risk-Taking Behavior Is Driven by Tolerance to Ambiguity," *Proceedings of the National Academy of Sciences of the United States of America* 109, no. 42 (2012): 17135–17140.

3. T. U. Hauser et al., "Cognitive Flexibility in Adolescence: Neural and Behavioral Mechanisms of Reward Prediction Error Processing in Adaptive Decision Making during Development," *Neuroimage* 104 (2015): 347–354.

4. L. Eiland and R. D. Romeo, "Stress and the Developing Adolescent Brain," *Neuroscience* 249 (2013): 162–171.

5. W. James et al., *The Principles of Psychology*, vol. 1, no. 2 (London: Macmillan, 1890).

6. Michael L. Mack, Bradley C. Love, and Alison R. Preston, "Dynamic Updating of Hippocampal Object Representations Reflects New Conceptual Knowledge," *Proceedings of the National Academy of Sciences of the United States of America* 113, no. 46 (2016): 13203–13208, https://www.pnas.org/doi/full/10.1073/pnas.1614048113.

7. N. A. Farb et al., "Interoceptive Awareness of the Breath Preserves Attention and Language Networks amidst Widespread Cortical Deactivation: A Within-Participant Neuroimaging Study," *eNeuro* 10, no. 6 (2023).

8. Maïté Delafin, Michael Ford, and Jerry Draper-Rodi, "Interoceptive Sensibility in Professional Dancers Living with or without Pain: A Cross-Sectional Study," *Medical Problems of Performing Artists* 37, no. 1 (2022): 58–66, https://www.ingentaconnect.com/content/scimed/mppa/2022/00000037/00000001/art00008.

9. J. David Creswell et al., "Neural Correlates of Dispositional Mindfulness during Affect Labeling," *Psychosomatic Medicine* 69, no. 6 (2007): 560–565, https://doi.org/10.1097/PSY.0b013e3180f6171f.

10. M. D. Lieberman et al., "Putting Feelings into Words: Affect Labeling Disrupts Amygdala Activity to Affective Stimuli," *Psychological Science* 18 (2007): 421–428.

CHAPTER 10

1. Lily Martis, "7 Companies with Epic Wellness Programs," Monster Jobs, https://www.monster.com/career-advice/article/companies-good-wellness-programs.

2. Hans Kraus and Ruth P. Hirschland, "Muscular Fitness and Health," *Journal of Health, Physical Education, Recreation* 24, no. 10 (1953): 17–19. Their findings were also published as "Muscular Fitness and Orthopedic Disability," *New York State Journal of Medicine* 54 (1954): 212–215; and "Minimum Muscular Fitness Tests in School Children," *Research Quarterly* 25 (1954): 178–188.

3. Carolyn E. Barlow et al., "Cardiorespiratory Fitness and Long-Term Survival in 'Low-Risk' Adults," *Journal of the American Heart Association* 1, no. 4 (2012): e001354, https://doi.org/10.1161/JAHA.112.001354.

4. M. Foucault, "Technologies of the Self," in *Technologies of the Self: A Seminar with Michel Foucault*, vol. 18, edited by Luther H. Martin, Huck Gutman, and Patrick H. Hutton (Amherst: University of Massachusetts Press, 1988).

5. J. Montero-Marin et al., "School-Based Mindfulness Training in Early Adolescence: What Works, for Whom and How in the MYRIAD Trial?," *Evidence-Based Mental Health* (2022): 1–8, https://doi.org/10.1136/ebmental-2022-300439.

6. J. L. Andrews, "Evaluating the Effectiveness of a Universal eHealth School-Based Prevention Programme for Depression and Anxiety, and the Moderating Role of Friendship Network Characteristics," *Psychological Medicine* (2022): 1–10, https://doi.org/10.1017/S0033291722002033; Stephanie Mertens, "The Effectiveness of Nudging: A Meta-Analysis of Choice Architecture Interventions across Behavioral Domains," *Proceedings of the National Academy of Sciences of the United States of America* 119, no. 1 (2022): e2107346118, https://www.pnas.org/doi/10.1073/pnas.2107346118.

7. "Mental Health: A Young Persons' Perspective," https://donothing.uk/.

8. Shelly McKenzie, *Getting Physical: The Rise of Fitness Culture in America* (Lawrence: University Press of Kansas, 2013), 38.

9. M. J. Poulin et al., "Minding Your Own Business? Mindfulness Decreases Prosocial Behavior for People with Independent Self-Construals," *Psychological Science* 32, no. 11 (2021): 1699–1708, https://doi.org/10.1177/09567976211015184.

10. Henry Taylor, "Sometimes, Paying Attention Means We See the World Less Clearly," *Psyche*, June 16, 2021, https://psyche.co/ideas/sometimes-paying-attention-means-we-see-the-world-less-clearly.

11. Ellen Choi, “What Do People Mean When They Talk about Mindfulness?,” *Clinical Psychology Review* 89 (2021): 102085, https://doi.org/10.1016/j.cpr.2021.102085.

12. Vanessa M. Brown, “Humans Adaptively Resolve the Explore-Exploit Dilemma under Cognitive Constraints: Evidence from a Multi-Armed Bandit Task,” *Cognition* 229 (2022), https://doi.org/10.1016/j.cognition.2022.105233.

13. Madhuleena Roy Chowdhury, “What Is the Mental Health Continuum Model?,” *Positive Psychology*, September 1, 2019, https://positivepsychology.com/mental-health-continuum-model/.

14. I. M. Balodis et al., “Neurofunctional Reward Processing Changes in Cocaine Dependence during Recovery,” *Neuropsychopharmacology* 41, no. 8 (2016): 2112–2121.

15. N. D. Volkow et al., “Loss of Dopamine Transporters in Methamphetamine Abusers Recovers with Protracted Abstinence,” *Journal of Neuroscience* 21, no. 23 (2001): 9414–9418.

16. Rachel E. Greenspan, “Meet the Man Who Popularized the Viral #Trashtag Challenge Getting People Around the World Cleaning Up,” *Time*, March 12, 2019, https://time.com/5549019/trashtag-interview/.

17. Drici Tani Younes, Facebook, https://www.facebook.com/photo?fbid=533669101448516&set=ecnf.100044162917667.

18. You track the work of others at www.trashtag.org or document your daily acts of care on www.cleanups.org.

Index

About the Authors

Norman Farb, PhD, is an associate professor of psychology at the University of Toronto Mississauga. He studies the psychology of well-being, focusing on the neuroscience of sensation and mental habits, such as how we think about ourselves and interpret our emotions. Farb is particularly interested in why people differ in their resilience to stress, depression, and anxiety. He lives in Toronto with his wife and two children. For sense foraging, Farb tries out new recipes, practices yoga, and cycles.

Zindel Segal, PhD, is Distinguished Professor of Psychology in Mood Disorders at the University of Toronto Scarborough and a cofounder of mindfulness-based cognitive therapy. His research on relapse vulnerability in affective disorder provided an empirical rationale for offering training in mindfulness meditation to those living with recurrent depression. Segal continues to advocate for the relevance of mindfulness-based clinical care in psychiatry and mental health. He lives in Toronto with his wife and three children. For sense foraging, Segal plays guitar, meditates in the morning, and goes on nature walks.

Farb and Segal first met more than fifteen years ago to work on a neuroimaging study of mindfulness meditation while Farb was a PhD student and Segal a professor of psychiatry in Toronto. Over the years they have bonded through their shared research and love of basketball. This is their first book together.